Quotes & Quips
for the Military Professional

CLOUGH

Compiled by: Bronston Clough
Cover by: Matthew Dail
Layout by: Preston O. Fitzgerald

©2013 Mentor Enterprises, Inc. All Rights Reserved.
ISBN-13: 978-0-9839714-4-3

1st Edition

Published by:
MENTOR ENTERPRISES, INC.

121 Castle Dr. Suite F
Madison, AL 35758
256.830.8282
info@mentorenterprisesinc.com

Subjects

Leadership	1
Motivational	29
Training	50
War & Combat	67
Character	97
Personal Courage	108
Patriotic	128
Discipline	145
Loyalty	156
Tactics	161
Integrity	171
Selfless Service	178
Teamwork	182
Morale & Esprit de Corps	192
Special Operations	203
Humorous	214
Scriptures	227
Infamous Leaders	231
Index	251

Dedicated to my mom, Pamela Clough, to whom I owe so much. Thank you for leading me to Christ and making me into the man I am today. I can't thank you enough and it is only fitting that a quote book is dedicated to you since the popular quote you used to tell me all the time when I was growing up still rings in my ears: *"I don't care how big you get, me and a baseball bat will always be bigger."* I love you!

Leadership

There is no limit to the good you can do if you don't care who gets the credit.

General George C. Marshall

A leader is a man who had the ability to get other people to do what they don't want to do, and like it.

Harry S. Truman

There is no type of human endeavor where it is so important that the leader understands all phases of his job as that of the profession of arms.

Major General James C. Fry

As we look ahead into the next century, leaders will be those who empower others.

Bill Gates

The threat of committing your force is a political tool to achieve your goal. Once you have committed your forces you must win because the enemy has nothing else to fear from you.

Anonymous

A leader takes people where they want to go. A great leader takes people where they don't necessarily want to go, but ought to be.

Rosalynn Carter, former First Lady

Leadership is the capacity to transform vision into reality.

Warren G. Bennis

If you can't be on time be early.

CSM (R) Leo Arsenault

Be an example to your men, in your duty and in private life. Never spare yourself, and let the troops see that you don't in your endurance of fatigue and privation. Always be tactful and well-mannered and teach your subordinates to do the same. Avoid excessive sharpness or harshness of voice, which usually indicates the man who has shortcomings of his own to hide.
Field Marshall Erwin Rommel

Innovation distinguishes between a leader and a follower.
Steve Jobs

Let your leaders lead.
Bronston Clough

Never forget that no military leader has ever become great without audacity. If the leader is filled with high ambition and if he pursues his aims with audacity and strength of will, he will reach them in spite of all obstacles.
MG Karl von Clausewitz

You don't lead by hitting people over the head-that's assault, not leadership.
General Dwight D. Eisenhower

I have to follow them. I am their leader.
Alexandre-Auguste Ledru-Rollin, Leader of the French Revolution of 1848

The final test of a leader is that he leaves behind him in other men the conviction and the will to carry on.
Walter Lippmann, 1889-1974

Reason and calm judgment, the qualities specially belonging to a leader.
Tacitus, 55-177

You must love Soldiers in order to understand them, and understand them in order to lead them.
Henri Turenne

No leader should put troops into the field merely to gratify his own spleen; no leader should fight a battle simply out of pique. But a kingdom that has once been destroyed can never come again into being; nor can the dead ever be brought back to life. Hence the enlightened leader is heedful, and the good leader full of caution.
Sun Tzu

A leader is a man who can adapt principles to circumstances.
General George S. Patton

A good leader must be a good follower.
Anonymous

Lead me, follow me, or get out of my way.
General George Patton

The mothers and fathers of America will give you their sons and daughters...with the confidence in you that you will not needlessly waste their lives. And you dare not. That's the burden the mantle of leadership places upon you. You could be the person who gives the orders that will bring about the deaths of thousands and thousands of young men and women. It is an awesome responsibility. You cannot fail. You dare not fail...

General H. Norman Schwarzkopf

Never tell people how to do things. Tell them what to do and they will surprise you with their ingenuity.

General George Patton Jr

Hog the blame and share the glory

Anonymous

A general is just as good or just as bad as the troops under his command make him.

General Douglas MacArthur

The real leader has no need to lead—he is content to point the way.
Henry Miller

Not the cry, but the flight of a wild duck leads the flock to fly and follow.
Chinese Proverb

Go to the people. Learn from them. Live with them. Start with what they know. Build with what they have. The best of leaders when the job is done, when the task is accomplished, the people will say we have done it ourselves.
Lao Tzu

A leader is a dealer in hope.
Napoleon Bonaparte

He who has never learned to obey cannot be a good commander.
Aristotle

A leader should encourage the members of his staff to speak up if they think the commander is wrong. He should invite constructive criticism.
GEN Omar Bradley

The more a leader is in the habit of demanding from his men, the surer he will be that his demands will be answered.
Karl von Clausewitz 1832

It is a grave error for the leader to surround himself with a 'yes' staff.
GEN Omar Bradley

An army of deer led by a lion is more to be feared than an army of lions led by a deer.
Chabrias 410–375 B.C.

Next is the commander. He must be smart, trustworthy, caring, brave, and strict.
Sun Tzu

Good leadership promotes professionalism—a renaissance of standards, involving quality of life, service, discipline and total commitment to our Army and the United States of America.
MG Albert Akers

Soldiers will not follow any battle leader with confidence unless they know that he will require full performance of duty from every member of the team.
GEN Dwight Eisenhower

Delegation of sufficient authority and proper use of subordinates helps develop future leaders. This is a moral responsibility of every commander.
LTC Stanley Bonta

The first quality of a commander is a cool head.
Napoleon Bonaparte

Authority is a poor substitute for leadership.
John Luther

There must be, within our Army, a sense of purpose. There must be a willingness to march a little farther, to carry a heavier load, to step out into the dark and the unknown for the safety and well-being of others.
GEN Creighton Abrams

The greatest leader in the world could never win a campaign unless he understood the men he had to lead.
GEN Omar Bradley

Fairness, diligence, sound preparation, professional skill and loyalty are the marks of American military leadership.
GEN Omar Bradley

Leadership is intangible, and therefore no weapon ever designed can replace it.
GEN Omar Bradley

Regardless of age or grade, Soldiers should be treated as mature individuals. They are men engaged in an honorable profession and deserve to be treated as such.
GEN Bruce Clarke

I know of no better reputation for an officer or noncommissioned officer to have with his men than he is a good manager and does not waste his Soldiers' time.
GEN Bruce Clarke

Leadership needed in ordinary units is different from that required by elite units.
GEN Bruce Clarke

A good leader can't get too far ahead of his followers.
Franklin Roosevelt

Men think as their leaders think.
Charles P. Summerall

Act before there is a problem. Bring order before there is disorder.
Taoism

There is no stigma attached to recognizing a bad decision in time to install a better one.
Lawrence J. Peter

Never discourage anyone who continually makes progress, no matter how slow.
Plato

Good management consists in showing average people how to do the work of superior people.
John Rockefeller

Handle an employee's problem swiftly, and as seriously as if it were a problem with your own paycheck.
Unknown

Responsibility educates.
Wendell Phillips

If your subordinates are not making an occasional mistake or two, it's a sure sign they're playing it too safe.
Tom Peters

Fail to honor your people and they will fail to honor you.
Eric Clark

We don't want geniuses for managers. We want people who can motivate other people to do good work.
Lee Iacocca

History will show that no man rose to military greatness, who could not convince his troops that he put them first.
General Maxwell Taylor

The badge of rank worn by leaders is a symbol of servitude—servitude to Soldiers.
GEN Maxwell Taylor

Any commander who fails to obtain his objective, and who is not dead or severely wounded, has not done his full duty.
GEN George Patton

60% of the art of command is the ability to anticipate.
SLA Marshall

The commander's will must rest on iron faith—faith in God, in his cause, and in himself.
MAJ Carrell Burnett

I don't want to hear about all the labor pains; just show me the baby.
Bronston Clough

The man who commands efficiently must have obeyed others in the past.
Cicero

A good leader takes a little more than his share of blame; a little less than his share of credit.
Arnold Glasgow

Example is not the main thing in influencing others. It is the only thing.
Albert Schweitzer

The climax of leadership is to know when to do what.
John Scotford

You will never be a leader unless you first learn to follow and be led.
Tiorio

The speed of the boss is the speed of the team.
Lee Iacocca

When the leader compromises, the whole organization compromises.
Charles Knight, CEO

A few men had the stuff of leadership in them, they were like rafts to which all the rest of humanity clung for support and hope.
Lord Moran

No leader, however great, can long continue unless he wins victories.
Field Marshall Viscount Montgomery

Leadership is a potent combination of strategy and character. But if you must be without one, be without the strategy.
GEN Norman Schwarzkopf

If I were asked to define leadership, I should say it is the projection of personality. It is the most intensely personal thing in the world because it is just plain you.
Field Marshall Sir William Slim

The qualities that distinguish a leader from other men are courage, will power, initiative, and knowledge. If you have not got those qualities you will not make a leader; if you have them, you will.
Field Marshall Sir William Slim

The beginning of leadership is a battle for the hearts and minds of men.
Field Marshall Bernard Montgomery

Leadership is the practical application of character.
R.E. Meinertzhagen

Few people are born leaders. Leadership is achieved by ability, alertness, experience and keeping posted; by willingness to accept responsibility; a knack of getting along with people; an open mind; and a head that stays clear under stress.
Franklin Field

The best executive is the one who has sense enough to pick good men to do what he wants done, and self-restraint enough to keep from meddling with them while they do it.
Teddy Roosevelt

Education is the mother of leadership.
Wendell Willkie

The man who commands efficiently must have obeyed others in the past.
Cicero

Where there is no vision, the people perish.
Solomon

When the crunch comes, your subordinates prefer to work with forthright plodders than with devious geniuses.
ADM James Stockdale

A good leader takes a little more than his share of blame; a little less than his share of credit.
Arnold Glasgow

You cannot antagonize and influence at the same time.
J.S. Knox

The boss fixes the blame for the breakdown. The leader fixes the breakdown. The boss says 'go,' the leader says 'let's go.'
H. Gordon Selfridge

The hallmark of leadership is to act upon convictions based on principles, rather than to pander to the whims and fancies of the mob.
Anwar Ibraham

Leadership is action, not position.
Donald H. McGannon

Men do well only what their leader checks.
GEN Bruce Clarke

Before making criticism, think through three things. Is it true? Is it kind? Is it necessary?
Bryan Gardner

Know your men.
Know your business.
Know yourself.
MAJ C.A. Bach

A good man obtains the confidence of those under him before putting burdens upon them; and the confidence of those above him before criticizing them.
Confucius

One of the tests of leadership is the ability to recognize a problem before it becomes an emergency.
Arnold Glasgow

The large caliber executive welcomes suggestions; the small-caliber executive resents them, imagining that he knows it all and that it is presumptuous for anyone to offer him advice.
B.C. Forbes

Talks with your subordinates are meant to be kept private. If you ever use their words against them in public, even in jest, you run the risk of never getting a straight answer from them again.
Note to P.Q. Carter

That which gets recognized and reinforced gets done. That which doesn't get recognized and reinforced doesn't get done.
The Greatest Management Principle in the World

We always have time for the things we put first.
Steven Covey

You need to know what is going on and what each team member does (and is capable of) before you make major changes in the group.
James Autry

The commander cannot shoot side by side with his Soldier- he must retain perspective.
David Wolffsohn

A great man leaves clean work behind him, and requires no sweeper up of the chips.
Elizabeth Barrett Browning

A good manager is a man who isn't worried about his own career, but rather the careers of those who work for him.
H.M.S. Burns

Leaders don't create followers, they create more leaders.
Tom Peters

There is no greater or more satisfying reward than that which comes from discovering and developing men.
A.A. Stambaugh

In the simplest terms, a leader is one who knows where he wants to go, and gets up, and goes.
John Erskine

People often think that high intelligence is a prerequisite. I don't believe so. I think a big heart and a strong spine are more important.
Elias Zerhouni

The buck stops here.
Harry Truman

When he was asked how someone might most surely earn people's esteem, he replied 'By the best words and finest actions.'
Plutarch

To be a successful leader, a commander must not only gain contact with his men, but he must give them the impression of fearlessness.
MG John Lejeune

The truly great leader overcomes all difficulties, and campaigns and battles are nothing but a long series of difficulties to overcome.
GEN George C. Marshall

Every great leader I have known has been a great teacher, able to give those around him a sense of perspective and to set the moral, social, and motivational climate among his followers.
ADM James Stockdale

A leader establishes the ethical...environment in which the entire operation is going to be accomplished.
GEN H. Norman Schwarzkopf

There is no command without leadership.
Sun Bin

We never lost sight of the reality that people, particularly gifted commanders, are what make units succeed. The way I like to put it, leadership is the art of accomplishing more than the science of management says is possible.
GEN Colin Powell

The greatest commander is he whose intuitions most nearly happen.
T.E. Lawrence

If your actions inspire others to dream more, learn more, do more and become more, you are a leader.
John Quincy Adams

High sentiments always win in the end, the leaders who offer blood, toil, tears and sweat always get more out of their followers than those who offer safety and a good time. When it comes to the pinch, human beings are heroic.
George Orwell

The led must not be compelled; they must be able to choose their own leader.
Albert Einstein

A leader is one who sees more than others see, who sees farther than others see, and who sees before others see.
Leroy Eimes

Leadership involves finding a parade and getting in front of it.
John Naisbitt

The task of the leader is to get his people from where they are to where they have not been.
Henry Kissinger

Great leaders are almost always great simplifiers, who can cut through argument, debate, and doubt to offer a solution everybody can understand.
GEN Colin Powell

People ask the difference between a leader and a boss. The leader leads and the boss drives.
Teddy Roosevelt

Our chief want is someone who can inspire us to be what we know we could be.
Ralph Waldo Emerson

Those who try to lead the people can only do so by following the mob.
Oscar Wilde

The task of leadership is not to put greatness into humanity, but to elicit it, for the greatness is already there.
John Buchan

Speak softly and carry a big stick.
Teddy Roosevelt

Reason and calm judgment, the qualities specially belonging to a leader.
Tacitus

I have just taken on a great responsibility. I will do my utmost to meet it.
ADM Chester Nimitz

You have the soul of a lion and the heart of a woman.
GEN Sickel about Chamberlain

I cannot consent to be feasting while my poor Soldiers are nearly starving.
GEN Robert E. Lee

You should never forget the power of example.
MG John Lejeune

Motivational

A man does not have himself killed for a half-pence a day or for a petty distinction. You must speak to the soul in order to electrify him

Napoleon Bonaparte

Also remember that in any man's dark hour, a pat on the back and an earnest handclasp may work a small miracle

Brigadier-General S.L.A Marshall, The Armed Forces Officer 1950

Regard your Soldiers as your children, and they will follow you into the deepest valleys. Look on them as your own beloved sons, and they will stand by you even unto death!

Sun Tzu

Men who think that their officer recognizes them are keener to be seen doing something honorable and more desirous of avoiding disgrace.

Xenophon, Greek historian (c. 430-355 BC)

There's no security on this earth, only opportunity.
General Douglas MacArthur

Age wrinkles the body. Quitting wrinkles the soul
General Douglas MacArthur

Success is how high you bounce when you hit bottom.
General George Patton Jr

I am a Soldier, I fight where I am told, and I win where I fight.
General George Patton Jr

We few, we happy few, we band of brothers. For he today that sheds his blood with me, Shall be my brother; be ne'er so vile, This day shall gentle his condition. And gentlemen in England now abed, Shall think themselves accursed they were not here, And hold their manhood's cheap whiles any speaks, That fought with us upon Saint Crispin's day.
William Shakespeare (King Henry V)

From time to time, the tree of liberty must be watered with the blood of tyrants and patriots.
Thomas Jefferson

It was a high counsel that I once heard given to a young person, Always do what you are afraid to do.
Ralph Waldo Emerson

A soldier will fight long and hard for a bit of colored ribbon.
Napoleon Bonaparte

Storms make oaks take roots.
Proverb

The only way of finding the limits of the possible is by going beyond them into the impossible.
Arthur C. Clarke

Without inspiration the best powers of the mind remain dormant. There is a fuel in us which needs to be ignited with sparks.
Johann Gottfried Von Herder

Work spares us from three evils: boredom, vice, and need.
Voltaire

Men's best successes come after their disappointments.
Henry Ward Beecher

You cannot plough a field by turning it over in your mind.
Unknown

Do not wait to strike till the iron is hot; but make it hot by striking.
William B. Sprague

Nothing will ever be attempted if all possible objections must first be overcome.

Samuel Johnson

It is not the critic who counts, not the one who points out how the strong man stumbled or how the doer of deeds might have done them better. The credit belongs to the man who is actually in the arena, whose face is marred with sweat and dust and blood; who strives valiantly; who errs and comes short again and again; who knows the great enthusiasms, the great devotions, and spends himself in a worthy cause; who, if he wins, knows the triumph of high achievement; and who, if he fails, at least fails while daring greatly, so that his place shall never be with those cold and timid souls who know neither victory nor defeat.

Teddy Roosevelt

Don't say it's impossible! Turn your command over to the next officer. If he can't do it, I'll find someone who can, even if I have to take him from the ranks!

General Thomas Stonewall Jackson

To the world you are someone, but to someone you are the world.
Unknown

The bigger they are, the harder they fall.
James King

It's not the size of the dog in the fight, but the size of the fight in the dog.
Mark Twain

The only way of finding the limits of the possible is by going beyond them into the impossible.
Arthur C. Clarke

We are what we repeatedly do. Excellence, therefore, is not an act but a habit.
Aristotle

We are still masters of our fate.
We are still captains of our souls.
Prime Minister Winston Churchill

Every artist was first an amateur.
Ralph Waldo Emerson

The more difficulties one has to encounter, within and without, the more significant and the higher in inspiration his life will be.
Horace Bushnell

No great man ever complains of want of opportunities.
Ralph Waldo Emerson

No one would have crossed the ocean if he could have gotten off the ship in a storm.
Bryan Gardner

He that cannot endure the bad will not live to see the good.
Jewish Proverb

Rest? Rest is for the dead.
Thomas Carlyle

Quitters never win; winners never quit.
Virginia Hutchinson

Nothing is invented and perfected at the same time.
John Ray

With ordinary talent and extraordinary perseverance, all things are attainable.
Thomas Buxton

Some men can get results if kindly encouraged, but give me the kinds that do things in spite of hell.
Elbert Hubbard

Great works are performed not by strength, but by perseverance.
Samuel Johnson

Every noble work was at first impossible.
Thomas Carlyle

The harder the conflict, the more glorious the triumph. What we obtain too cheaply, we esteem too lightly.
Thomas Paine

The greatest achievement is not in never falling, but in rising again after you fall.
Vince Lombardi

Never despair, but if you do, work on in despair.
Edmund Burke

The first and best victory is to conquer self.
Plato

To become a champion, fight one more round.
James Corbett

Things may come to those who wait, but only those things left behind by those who hustle.
Abraham Lincoln

Don't be afraid to take a big step when one is indicated. You can't cross a chasm in two small jumps.
David Lloyd George

All glory comes from daring to begin.
Eugene F. Ware

Action may not bring happiness, but there is no happiness without action
Benjamin Disraeli

'I must do something' will solve more problems than 'something must be done.'
Bruce Barton

You may be disappointed if you fail, but you are doomed if you don't even try.
Beverly Sills

One today is worth two tomorrows.
Francis Quarles

I may lose a battle, but I will lose no time.
Napoleon Bonaparte

Better three hours too soon than one minute too late.
William Shakespeare

One acre of action is worth the whole land of promise.
Rudyard Kipling

Sometimes we take credit for being patient when we are only putting off doing something unpleasant. Live neither in the past nor in the future, but let each day's work absorb all your interest, energy and enthusiasm. The best preparation for tomorrow is to do today's work superbly well.
Sir William Osler

Action may not always bring happiness; but there is no happiness without action.
Benjamin Disraeli

When a man boasts of what he'll do tomorrow we should like to find out what he did yesterday.
Benjamin Franklin

It is better to undertake a large task and get it half done then to undertake nothing and get it all done.
W. Marshall Craig

If you have a number of disagreeable duties to perform, always do the most disagreeable one first.
Josiah Quincy

Today only is thine. If thou procrastinate , thou loseth. Which lost, is lost forever.
Francis Quarles

Nothing would be done at all if a man waited until he could do it so well that no-one could find fault with it.
Cardinal John Henry Newman

The world stands aside and lets anyone pass who knows where he is going.
David Gordon

The world is most blessed by men who do things, and not by men who merely talk about them.
James Oliver

If there is no wind, row.
Latin Proverb

It is better to err on the side of initiative than inactivity.
B.C. Forbes

The time to repair the roof is when the sun is shining.
John F. Kennedy

Big jobs usually go to the men who have proved their ability to outgrow the small ones.
Ralph Waldo Emerson

From a little spark may burst a mighty flame.
Dante Aleghieri

Success is not so much achievement as achieving. Refuse to join the crowd that plays not to lose; play to win.
David Mahoney

You show me a gracious loser and I'll show you a failure.
Unknown

Triumph usually comes from putting a little more 'umph' into your 'try.'
Howard Crimson

I firmly believe that any man's finest hour, his greatest fulfillment to all he holds dear, is the moment when he has worked his heart out in a good cause and lies exhausted, but victorious, on the field of battle.
Vince Lombardi

Follow through: stopping at third base adds no more to the score than striking out.
Alexander Animator

It isn't enough to put your best foot forward, you have to follow through with the other foot.
Jacob Braude

The joy of living comes from immersion in something that we know to be bigger, better, more enduring and worthier than we are.
John Mason Brown

Always bear in mind that your own resolution to succeed is more important than any one thing.
Abraham Lincoln

If you have accomplished all that you have planned for yourself, you have not planned enough.
Edward Hale

Show me a man who doesn't know the meaning of the word 'fail' and I'll show you a man who ought to buy a dictionary.
Albert Einstein

I have generally found that a man who is good at an excuse is good at nothing else.
Benjamin Franklin

The secret of greatness is simple: Do better work than anyone in your field—and keep on doing it.
Wilfred Peterson

The man who has the will to undergo all labor may achieve any goal.
Menander

You become strong by defying defeat, and by turning loss to gain and failure to success.
Napoleon Bonaparte

The greater the obstacle, the more glory in overcoming it.
Moliere

We will either find a way or make one.
Hannibal

Great spirits have always encountered violent opposition from mediocre minds.
Albert Einstein

Self-confidence is the first requisite to great undertakings.
Samuel Johnson

We cannot become what we need to be by remaining what we are.
Max DuPree

Focus on where you want to go, not what you fear.
Anthony Robbins

You must have long range goals to keep you from being frustrated by short range failures.
Charles Noble

Many of life's failures are people who did not realize how close they were to success when they gave up.
Thomas Edison

Do not despise your situation; in it you must act, suffer and conquer. From every point on earth we are equally near to heaven and the infinite.
Henri Amiel

There are too many people praying for the mountains of difficulty to be removed, when what they really need is courage to climb them.
Raili Jeffery

Endure and persist. The pain will do you good.
Ovid

I've found that it's not good to talk about your troubles. Eighty percent of the people who hear them don't care, and the other twenty percent are glad you're having trouble.
Tommy Lasorda

A happy person is not a person in a certain set of circumstances, but rather a person with a certain set of attitudes.
Hugh Downs

If you have built castles in the air, your work need not be lost. Now go, and put the foundations under them.
Henry David Thoreau

When you are an anvil, be patient. When you are the hammer, strike.
Arabian Proverb

The trouble with an opportunity is that it always comes disguised as hard work.
Will Rogers

Never mind your happiness. Do your duty.
Will Durant

To every man there comes in his lifetime that special moment where he is physically tapped on the shoulder and offered the chance to do a very special thing, unique to him and fitted to his talent; what a tragedy if that moment finds him unprepared or unqualified for the work which would be his finest hour.
Prime Minister Winston Churchill

Great men are rarely known to fail in their most perilous enterprises...Is it because they are lucky that they became great? No, but being great, they have been able to master luck.
Napoleon Bonaparte

My business is to succeed, and I'm good at it. I create my Iliad by my actions, create it day by day.
Napoleon Bonaparte

This is a long tough road we have to travel. The men that can do things are going to be sought out just as surely as the sun rises in the morning.
GEN Dwight Eisenhower

There is nothing impossible! Give your orders, support them with firmness, and you will see every obstacle vanish.
LTG Johann von Ewald

I stand at the altar of murdered men and while I live I fight their cause.
Florence Nightingale

Give me a thousand men crazy enough to conquer hell, and we will conquer hell.
Solon

Training

The sergeant is the Army.

GEN Dwight Eisenhower

The hardships of forced marches are often more painful than the dangers of battle.

General Thomas Stonewall Jackson

Endure and persist. The pain will do you good.

Ovid

Technical training is important, but it accounts for less than 20% of one's success. More than 80% is due to the development of one's personal qualities, such as initiative, thoroughness, concentration, decision, adaptability, organizing ability, observation industry, and leadership.

Dr. G.P. Koch

The human capacity is incredible: We can adapt to anything if we make the right demands upon ourselves incrementally.
Anthony Robbins

The best form of taking care of troops is first-class training, for this saves unnecessary casualties.
Erwin Rommel

The ultimate measure of a man is not where he stands in moments of comfort, but where he stands at times of challenge and controversy.
Martin Luther King, Jr.

This is War. It is the most important skill in the nation. It is the basis of life and death. It is the philosophy of survival or destruction. You must know it well.
Sun Tzu

We aren't going to try to train you, we're going to try to kill you.
Soldier I, SAS

In no other profession are the penalties for employing untrained personnel so appalling or so irrevocable as in the military.
Douglas MacArthur

Si Vis Pacem, Para Bellum.
(If you want peace, prepare for war)
Flavius Vegetius Renatus. Roman Military strategist. c. 390. A.D

We must remember that one man is much the same as another, and that he is best who is trained in the severest school.
Thucydides, History of the Peloponnesian War (431-404 B.C.)

On the fields of friendly strife are sown the seed that on other days and other fields will bear the fruits of victory.
General Douglas MacArthur

A pint of sweat will save a gallon of blood.
General George S. Patton, Jr

Good units do the routine things, routinely.
BG Chris Hughes

You owe it to your men to require standards which are for their benefit even though they may not be popular at the moment.
GEN Bruce Clarke

Don't begrudge the time you spend developing, coaching and helping your people to grow so they can carry on when you're gone. It's one of the best signs of good leadership.
Bernard Baruch

I yield to no man in sympathy for the gallant men under my command; but I am obliged to sweat them tonight, so that I may save their blood tomorrow.
General Thomas Stonewall Jackson

Make your plans to fit the circumstances.
General George S. Patton, Jr

The aim of military training is not just to prepare men for battle, but to make them long for it.
Louis Simpson

Success demands a high level of logistical and organizational competence.
General George S. Patton, Jr

Few men are born brave; many become so through training and force of discipline.
Flavius Renatus

To be prepared for war is one of the most effective ways of preserving peace.
George Washington

The instruments of battle are only valuable if one knows how to use them.
Ardant Du Picq

Put your trust in God, but keep your powder dry.
Oliver Cromwell

Let our advance worrying become advance thinking and planning.
Prime Minister Winston Churchill

Practice does not make perfect. Perfect practice makes perfect.
Vince Lombardi

A good scare is often worth more than good advice.
Edgar Watson Howe

Train hard, fight easy, and win. Train easy, fight hard, and die.
Unknown

There are no secrets to success. It is a matter of preparation, hard work, and learning from failure.
GEN Colin Powell

Though all under heaven be at peace, if the art of war be forgotten there is peril.
Chinese proverb

I would caution you always to remember that an essential qualification of a good leader is the ability to recognize, select, and train junior leaders.
GEN Omar Bradley

To lead an untrained people to war is to throw them away.
Confucius c500 BC

If we exercise now, how much stronger we will be when the test comes.
William James

Military history is the most effective way of training during peacetime.
Von Maltke

Excellence doesn't happen by accident. It takes preparation, teamwork, consistency, and dedication.
Christopher Lewis

We are what we repeatedly do. Excellence, therefore, is not an act, but a habit.
Aristotle

The will to win means nothing without the will to prepare.
Juma Ikanga'a (Olympic Runner)

Luck is when preparation meets opportunity.
Sterling Sill

Most spectacular performances are preceded by a great deal of unspectacular preparation.
Unknown

There is no studying on the battlefield. It is then simply a case of doing what is possible and making use of what one knows. In order to make a little possible one must know much.
Ferdinand Foch

What we will do on some great occasion will probably depend on what we already are; and what we are will be the result of previous years of self discipline.
Henry Louis Liddon

He, therefore, who desires peace should prepare for war. He who aspires to victory should spare no pains to form his Soldiers. And he who hopes for success should fight on principle, not chance.
Vegetius

How much effort it takes to make things look effortless.
Jacqueline Kennedy Onassis

A good scare if often worth more than good advice.
Edgar Watson Howe

One often learns more from ten days of agony than from ten years of contentment.
Merle Shain

Your performance depends on your people. Select the best, train them and back them.
Donald Rumsfeld

Reject your sense of injury and the injury itself disappears.
Emperor Marcus Aurelius

Press on. Obstacles are seldom the same size tomorrow as they are today.
Robert H. Schuller

Two things a leader is supposed to do: lead Soldiers and units during the battle; prepare Soldiers and units to fight the battle.
COL Mike Malone

To win a war quickly takes long preparation.
Latin Proverb

Form good habits. They're as hard to break as the bad ones.
Army NCO

Slow is smooth and smooth is fast.
Army saying

The Lacedaemonians had learned that true safety was to be found in long previous training, and not in eloquent exhortations uttered when they were going into action.
Thucydides

Training is light, and lack of training is darkness.
Field Marshall Prince Aleksandr V

Every officer and Soldier who is able to do duty ought to be busily engaged in military preparation, by hard drilling.
LTG Stonewall Jackson

More important even than military education is practical military training.
MG John Lejeune

In war the issues are decided not by isolated acts of heroism but by the general training and spirit of an army.
Field Marshall Albert Kesselring

Perhaps somewhere in the primal reaches of our Army's memory, left over from the days ten thousand years ago when armies first began, there's a simple and fundamental formula: SKILL + WILL= KILL.
COL Dandridge Malone

People die because they are incapable and are defeated because they are unsuitable. Therefore the most important thing in handling troops is to train them.
Wu Ch'i

The troops should be exercised frequently, cavalry as well as infantry, and the general should often be present to praise some, to criticize others, and to see with his own eyes that the orders...are observed exactly.
Frederick the Great

The Soldiers like training provided it is carried out sensibly.
Field Marshall Prince Aleksandr V

The unfortunate thing is that so many commanders don't recognize dull training. But their troops do.
LTG Arthur Collins

Body and spirit I surrendered whole to harsh instructors- and received a soul.
Rudyard Kipling

The first day I was at camp I was afraid I was going to die. The next two weeks my sole fear was that I wasn't going to die. And after that I knew I'd never die because I'd become so hard nothing could kill me.
American Soldier quoted in Cowing, Dear Folks at Home, 1919

The whole training of an officer seeks to accomplish one purpose- to instill in him the ability to take over in battle in a time of crisis.
GEN Matthew Ridgway

One great difficulty of training the individual Soldier in peace is to instill discipline and yet to preserve the initiative and independence needed in war.
Field Marshall Viscount Wavell

More emphasis will be placed on the hardening of men and officers.
GEN George S. Patton

Truly then, it is killing men with kindness not to insist upon physical standards during training which will give them a maximum fitness for the extraordinary stresses of campaigning in war.
S.L.A. Marshall

Battles are fought by platoons and squads. Place emphasis on small unit combat instruction so that it is conducted with the same precision as close-order drill.
GEN George S. Patton

Make peace a time for training for war, and battle an exhibition of bravery.
The Emperor Maurice

Accustom yourself to tireless activity.
Field Marshall Prince Aleksandr V

Win with ability, not with numbers.
Field Marshall Prince Aleksandr V

In the moment of action remember the value of silence and order.
Phormio of Athens

Drill is necessary to make the Soldier steady and skillful, although it does not warrant exclusive attention.
Field Marshall Maurice Comte de Saxe

Drill may be beautiful: but beauty is not perceptible when you are expecting a punishment every moment for not doing it well enough. Dancing is beautiful because it's the same sort of thing, without the sergeant major.
T.E. Lawrence

Combat experience has proven that ceremonies, such as formal guard mounts, formal retreat formations, and regular supervised reveille formations, are a great help, and, in some cases, essential, to prepare men and officers for battle, to give them that perfect discipline, that smartness of appearance, that alertness without which battles cannot be won.
GEN George S. Patton

What can a Soldier do who charges when out of breath?
Flavius Vegetius Renatus

Will power, determination, mental poise, and muscle control all march hand-in-hand with the general health and well-being of the man.
BG S.L.A. Marshall

The civil comparison to war must be that of a game, a very rough and dirty game, for which a robust mind and body are essential.
Field Marshall Viscount Wavell

Drill your Soldiers well, and give them a pattern yourself.
Alexander V. Suvorov

So sensible were the Romans of the imperfections of valor without skill and practice that, in their language, the name of an Army, was borrowed from the word which signified exercise.
Edward Gibbon, Exercitus means Army in Latin

War & Combat

People sleep peaceably in their beds at night only because rough men stand ready to do violence on their behalf.

George Orwell

Hard pressed on my right. My center is yielding. Impossible to maneuver. Situation excellent. I am attacking.

Ferdinand Foch at the Battle of the Marne

I wish to have no connection with any ship that does not sail fast, for I intend to go in harm's way.

John Paul Jones

One officer who went to Sorrento tells this story: After going ashore he went looking for Darby. Approaching one member in a Ranger uniform, he asked his usual question, 'Do you know where I can find Col. Darby?' A slow grin crossed the face of the husky Soldier as he answered, You'll never find him this far back.
Unknown

Onward we stagger, and if the tanks come, may God help the tanks.
Col. William O. Darby

Battles are sometimes won by generals; wars are nearly always won by sergeants and privates.
F.E. Adcock, British classical scholar

He who stays on the defensive does not make war, he endures it
Field Marshal Colmar Baron von der Goltz, 1883

War is the mother of everything.
Heraclitus, Greek philosopher (535-475BC)

In peace sons bury fathers, but war violates the order of nature, and fathers bury sons.
Heroditus, greek historian c. 484-425 B.C.

It is well that war is so terrible, else we should grow too fond of it.
General Robert E. Lee

Wars may be fought with weapons, but they are won by men.
General George Patton Jr

Cry 'Havoc' and let slip the dogs of War.
William Shakespeare (Julius Caesar)

War is cruelty. There's no use trying to reform it, the crueler it is the sooner it will be over.
William Tecumseh Sherman

Many a boy here today looks on war as all glory, but boys war is all hell.
William Tecumseh Sherman

Never in the field of human conflict was so much owed by so many to so few.
Prime Minister Winston Churchill

No bastard ever won a war by dying for his country. He won it by making the other poor dumb bastard die for his country.
General George Patton Jr

The Nation that makes a great distinction between its scholars and its warriors will have its thinking done by cowards and its fighting done by fools.
Thucydides

In war, you win or lose, live or die - and the difference is an eyelash.
General Douglas MacArthur

In war there is no substitute for victory.
General Douglas MacArthur

One cannot wage war under present conditions without the support of public opinion, which is tremendously molded by the press and other forms of propaganda.
General Douglas MacArthur

We need to destroy not attack, not damage, not surround. I want to destroy the Republican Guard
General H. Norman Schwarzkopf 1991

Yesterday at the beginning of the ground war Iraq had the fourth largest army in the world. Today they have the second largest army in Iraq
Attributed to General H. Norman Schwarzkopf

Infantry must move forward to close with the enemy. It must shoot in order to move.... To halt under fire is folly. To halt under fire and not fire back is suicide. Officers must set the example
General George Patton Jr War as I knew it 1947

Fixed fortifications are monuments to the stupidity of man.
General George Patton Jr

If you are going to win any battle, you have to do one thing. You have to make the mind run the body. Never let the body tell the mind what to do... the body is never tired if the mind is not tired.
General George S. Patton

Just drive down that road, until you get blown up
General George Patton, about reconnaissance troops

Then, Sir, we will give them the bayonet!
Stonewall Jackson's reply to Colonel B.E Bee when he reported that the enemy was beating them back at the first battle of Bull Run, July 1861.

When war does come, my advice is to draw the sword and throw away the scabbard.
General Thomas Stonewall Jackson

Once you get them running, you stay right on top of them, and that way a small force can defeat a large one every time... Only thus can a weaker country cope with a stronger; it must make up in activity what it lacks in strength.
General Thomas Stonewall Jackson

My troops may fail to take a position, but are never driven from one!
General Thomas Stonewall Jackson

Violence, naked force, has settled more issues in history than has any other factor and the contrary opinion are wishful thinking at its worst. Breeds that forget this basic truth have always paid for it with their lives and freedoms.
Robert A. Heinlein

The cohesion that matters on the battlefield is that which is developed at the company, platoon and squad levels.
GEN E. C. Meyer

Small forces are not powerful; however, large forces cannot catch them.

Sun Tzu

War is an ugly thing but not the ugliest of things; the decayed and degraded state of moral and patriotic feelings which thinks that nothing is worth war is much worse. A man who has nothing for which he is willing to fight, nothing which is more important than his own personal safety, is a miserable creature and has no chance of being free unless made and kept so by the exertions of better men than himself.

John Stuart Mill

Let your plans be dark and as impenetrable as night, and when you move, fall like a thunderbolt.

Sun Tzu, The Art of War

There is only one tactical principle which is not subject to change. It is to use the means at hand to inflict the maximum amount of wound, death, and destruction on the enemy in the minimum amount of time.

General George S. Patton, Jr.

I have not yet begun to fight.
John Paul Jones (aboard the Bon Homme Richard), Sept. 1779

War is the continuation of policy (politics) by other means.
MG Karl von Clausewitz

I don't know whether war is an interlude during peace, or peace is an interlude during war.
Georges Clemenceau

The essence of war is violence. Moderation in war is imbecility
British Sea Lord John Fisher

So in war, the way is to avoid what is strong and to strike at what is weak.
Sun Tzu, The Art of War

The art of war is, in the last result, the art of keeping one's freedom of action
Xenophon, Greek historian (c. 430-355 BC)

Old Soldiers never die; they just fade away.
General Douglas MacArthur

May God have mercy upon my enemies, because I won't.
General George Patton Jr

It is the unconquerable nature of man and not the nature of the weapon he uses that ensures victory.
General George Patton Jr

Once we have a war there is only one thing to do. It must be won. For defeat brings worse things than any that can ever happen in war.
Ernest Hemingway

Under divine blessing, we must rely on the bayonet when firearms cannot be furnished
Stonewall Jackson 1861

It is fatal to enter any war without the will to win it.
General Douglas MacArthur

I am quite confident that in the foreseeable future armed conflict will not take the form of huge land armies facing each other across extended battle lines, as they did in World War I and World War II or, for that matter, as they would have if NATO had faced the Warsaw Pact on the field of battle.

General H. Norman Schwarzkopf

If we do go to war, psychological operations are going to be absolutely a critical, critical part of any campaign that we must get involved in.

General H. Norman Schwarzkopf

Few men are killed by the bayonet, many are scared by it. Bayonets should be fixed when the fire fight starts

General George Patton Jr, War as I knew it 1947

You may fly over a land forever; you may bomb it, atomize it, pulverize it and wipe it clean of life, but if you desire to defend it, protect it, and keep it for civilization, you must do this on the ground, the way the Roman legions did, by putting your young men into the mud.

T.R. Fehrenbach

There's no such thing as a crowded battlefield. Battlefields are lonely places.
LTG Alfred Gray

I have never advocated war except as a means of peace.
Ulysses S. Grant

We have met the enemy and they are ours!
Commodore Oliver Hazard Perry

The most persistent sound which reverberates through man's history is the beating of war drums.
Arthur Koesther

With two thousand years of examples behind us we have no excuse, when fighting, for not fighting well.
T.E. Lawrence

No art or science is as difficult as that of war.
Henry Lloyd

God fights on the side with the best artillery.
Napoleon Bonaparte

Where a goat can go, a man can go, where a man can go, he can drag a gun.
COL William Phillips 1777

Nothing is so exhilarating in life as to be shot at with no result.
Prime Minister Winston Churchill

The day of parachute troops is over.
Adolf Hitler after heavy losses at Crete in 1941

There are no atheists in foxholes.
LTC Warren Clear

Praise the Lord and pass the ammunition.
Chaplain Howell Forgy, Pearl Harbor

In the final choice a Soldier's pack is not so heavy a burden as a prisoner's chains.
GEN Dwight Eisenhower

Battles are won by the infantry, the armor, the artillery, and air teams, by Soldiers living in the rains and huddling in the snow. But wars are won by the great strength of a nation—the Soldier and the civilian working together.
GEN Omar Bradley

The most terrible job in warfare is to be a second lieutenant leading a platoon when you are on the battlefield.
GEN Dwight Eisenhower

He is what his home, his religion, his schooling, and the moral code and ideals of his society have made him. The Army cannot unmake him. It must reckon with the fact that he comes from a civilization in which aggression connected with the taking of life, is prohibited and unacceptable
BG S. L. A. Marshall

The God of war hates those who hesitate.
Euripedes

Everyone's a pacifist between wars. It's like being a vegetarian between meals.
Colman McCarthy

In Flanders fields the poppies grow Between the crosses, row on row, That mark our place, and in the sky, The larks, still bravely singing, fly, Scarce heard amid the guns below.
John McCrae

The tragedy of war is that it uses man's best to do man's worst.
Henry Fosdick

There is no room in war for delicate machinery.
Archibald Wavell

If, however, there is to be a war of nerves let us make sure our nerves are strong and are fortified by the deepest convictions of the heart.
Prime Minister Winston Churchill

War is a conflict of great interests which is settled by bloodshed, and only in that is it different from others.
MG Karl von Clausewitz

The bloody solution of the crisis, the effort for the destruction of the enemy's forces, is the firstborn son of war.
MG Karl von Clausewitz

Having to do it for the first time in combat is a chastening experience, it gives a man religion.
MG J.M. Gavin after his first combat jump, 1944

War hath no fury like a noncombatant.
Charles Edward Montague

Good logistics is combat power.
LTG William Pagonis

Intuition is often crucial in combat, and survivors learn not to ignore it.
COL F.F. Parry, USMC

When you get into combat, the only one you can trust is yourself and the man next to you.
SGT Garnier, 506th PIR, 101st

The difference between combat and sports is that in combat you bury the guy who comes in second.
Unknown

There is at least one thing worse than fighting with allies—And that is to fight without them
Prime Minister Winston Churchill

He who wishes to fight must first count the cost.
Sun Tzu

The harder the fighting and the longer the war, the more the infantry, and in fact all the arms, lean on the gunners.
Field Marshall Montgomery

It is a paradox to hope for victory without fighting. The goal of the man who makes war is to fight in the open field to win a victory.
Field Marshall Montecuccoli

Beware that, when fighting monsters, you yourself do not become a monster... for when you gaze long into the abyss, the abyss gazes also into you.
Friedrich Nietzsche

I don't care how they dress so long as they mind their fighting.
Sir Thomas Picton, Waterloo

Get up and get moving. Follow me!
MG Aubrey Newman 1944

A riot is a spontaneous outburst. A war is subject to advance planning.
Richard Nixon

In the moment of action remember the value of silence and order.
Phormio of Athens

It makes no difference what men think of war, said the judge. War endures. As well ask men what they think of stone. War was always here. Before man was, war waited for him. The ultimate trade awaiting the ultimate practitioner.
Cormac McCarthy

The legitimate object of war is a more perfect peace.
William Tecumseh Sherman

The essence of war is violence. Moderation in war is imbecility.
British Sea Lord John Fisher

War is the domain of friction, uncertainty, danger, chance, physical exertion, and suffering.
MG Karl von Clausewitz

It is never very crowded at the front.
GEN Creighton Abrams

If you start to take Vienna, take Vienna!
Napoleon Bonaparte

When you appeal to force, there's only one thing you must never do- lose.
Dwight Eisenhower

Battles are the principal milestones in secular history.
Prime Minister Winston Churchill

I have always regarded the forward edge of the battlefield as the most exclusive club in the world.
GEN Sir Brian Horrocks

When bayonets deliberate, power escapes from the hands of the government.
Napoleon Bonaparte

Rangers of Connaught! It is not my intention to expend any powder this evening. We'll do this business with cold steel.
GEN Sir Thomas Picton

What the American people want to do is fight a war without getting hurt. You cannot do that any more than you can go into a barroom brawl without getting hurt.
Chesty Puller

I love war and responsibility and excitement. Peace is going to be hell on me.
GEN George S. Patton

In war a Soldier must expect short commons, short sleep, and sore feet.
M.I. Dragomirov

The Infantry must ever be valued as the very foundation and nerve of an army.
Niccolo Machiavelli

Never mistreat the enemy by halves.
Prime Minister Winston Churchill

When the smoke cleared away, it was the man with the sword, or the crossbow, or the rifle, who settled the final issue on the field.
GEN George C. Marshall

I love the infantryman because they are the underdogs. They are the mud-rain-and-wind boys. They have no comforts, and they learn to live without the necessities. And in the end they are the guys that wars can't be won without.
Ernie Pyle

Look into an infantryman's eyes and you can tell how much war he has seen.
SGT Bill Mauldin

The art of war is, in the last result, the art of keeping one's freedom of action.
Xenophon

Christmas 1944 dawned clear and cold; lovely weather for killing Germans, although the thought seemed somewhat at variance with the spirit of the day.
GEN George S. Patton

Youths only want one thing, to kill you so they can go to paradise.
Osama bin Laden

The purpose of the military is to kill, and if you cannot stomach that, you should not have a military.
LTC Ralph Peters

It is bias to think that the art of war is just for killing people. It is not to kill people, it is to kill evil. It is a stratagem to give life to many people by killing the evil of one person.
Yagyu Munenori

War is an art and as such is not susceptible of explanation by fixed formula.
GEN George S. Patton

Forward, even with only a spear.
Samurai Proverb

We are so outnumbered there's only one thing to do. We must attack.
ADM Andrew Cunningham

I hope to God that I have fought my last battle.
The Duke of Wellington

I don't know what effect these men will have upon the enemy, but, by God, they terrify me.
The Duke of Wellington

The Soldier is the primary and most powerful mechanism of war.
Jose Vilabla, Spanish General

There are only two kinds of people on this beach; the dead and those about to die. So let's get the hell out'a here.
COL George Taylor, Omaha Beach, CDR 16th INF

War is cruelty. There's no use trying to reform it, the crueler it is the sooner it will be over.
GEN William T. Sherman

There's only one truth about war: people die.
Sheridan

It is not the big armies that win battles, it is the good ones!
Marshall Maurice de Saxe

Soldiers' bellies are not satisfied with empty promises and hopes.
Peter the Great

The bastards have never been bombed like they're going to be bombed this time.
Richard Nixon

Hold that ground at all hazards.
COL Vincent Strong to Joshua Chamberlain about Little Round Top

At times I saw around me more of the enemy than of my own men; gaps opening, swallowing, closing again, squads of stalwart men who had cut their way through us, disappearing as if translated. All around a strange, mingled roar.
COL Joshua Chamberlain

Unless history can teach us how to look at the future, the history of war is but a bloody romance.
MG J.F.C. Fuller

To make war without a thorough knowledge of the history of war is on a par with the casualness of a doctor who prescribes medicine without taking the trouble to study the history of the case he is treating.
Sir Basil Liddell Hart

Success in war, like charity in religion, covers a multitude of sins.
GEN Sir William Napier

What will history say- what will posterity think?
Napoeon

As in all battles the dead and woundest came chiefly from the best and the bravest.
Field Marshall Lord Carver

The business of a Soldier is to fight.
Stonewall Jackson

The acid test of battle brings out the pure metal.
GEN George S. Patton

It is warm work; and this day may be the last to any of us at a moment. But mark you! I would not be elsewhere for thousands.
ADM Viscount Nelson, Battle of Copenhagen

Boys, remember that it is my custom to sleep on the battlefield!
Napoleon Bonaparte

'Safety first' is the road to ruin in war.
Prime Minister Winston Churchill

After all, the most distressing and the most expensive thing in war is- to get men killed.
MG J.F.C. Fuller

The successful Soldier wins his battles cheaply so far as his own casualties are concerned, but he must remember that violent attacks, although costly at the time, save lives in the end.
GEN George S. Patton

I think it would be better to order up some artillery and defend the present location.
GEN Ulysses Grant

The unerring hand of providence shielded my men.
Andrew Jackson, Battle of New Orleans

As to the field of battle it is the abiding place of the dead. And he who decides to die will live, and he who wishes to live will die.
Wu C'hi

We have nothing left in the world but what we can win with our swords.
Hannibal

Go, therefore, to meet the foe with two objects before you, either victory or death. For men animated by such a spirit must always overcome their adversaries, since they go into battle ready to throw away their lives.
Scipio Africanus

If no one had the right to give his views on military operations except when he is frozen, or faint from heat and thirst, or depressed from privation and fatigue, objective and accurate views would be even rarer than they are.
MG Carl von Clausewitz

No human being knows how sweet sleep is but a Soldier.
COL John S. Mosby

With brave infantry and bold commanders mountain ranges can usually be forced.
LTG Antoine-Henri Baron de Jomini

No operation of war is more critical than a night march.
Prime Minister Winston Churchill

Character

You cannot dream yourself into a character; you must hammer and forge yourself one.

Henry David Thoreau

Be your character what it will, it will be known, and nobody will take it upon your word.

Lord Chesterfield

Reputation is what men and women think of us; character is what God and angels know of us.

Thomas Paine

The essential thing is not knowledge, but character.

Joseph Le Conte

Sow an act, and you reap a habit; sow a habit, and you reap a character; sow a character, and you reap a destiny.

George Dana Boardman

Our character is but the stamp on our souls of the free choices of good and evil we have made through life.
John C. Geikie

Reputation is for time; character is for eternity.
J. B. Gough

Character is a diamond that scratches every other stone.
Cyrus A. Bartol

The true sense of our character is not measured by what we do when times are good but what we do when times are tough.
CSM(R) Mark Gerecht

Character and personal force are the only investments that are worth anything.
Walt Whitman

Character is, in the long run, the decisive factor in the life of individuals and of nations alike.
Theodore Roosevelt

Moral courage is the most valuable and usually the most absent characteristic in men.
General George S. Patton, Jr

To sin by silence, when they should protest, makes cowards of men
Attributed to Abraham Lincoln

The ultimate measure of a man is not where he stands in moments of comfort and convenience, but where he stands at times of challenge and controversy.
Marin Luther King Jr.

It is, indeed, an observable fact that all leaders of men, whether as political figures, prophets, or Soldiers, all those who can get the best out of others, have always identified themselves with high ideals.
GEN Charles DeGaulle

God grant that men of principle shall be our principal men.
Thomas Jefferson

I would lay down my life for America, but I cannot trifle with my honor.
John Paul Jones

The untruthful Soldier trifles with the lives of his countrymen and the honor and safety of his country.
GEN Douglas MacArthur

Personal and professional excellence...you can't do one without the other...they are all wrapped up in the word 'character'.
GEN John Wickham, Jr.

A man's character is his fate.
Heraclitus

Ability may get a man to the top, but it takes character to keep him there.
Unknown

In matters of style, swim with the current; in matters of principle, stand like a rock.
Thomas Jefferson

Character is the ability to follow through with a commitment long after the original motivation has passed.
Larry Beckham

Never give up and never give in.
Hubert Humphrey

Leadership is the practical application of character.
R.E. Meinertzhagen

You can't dream yourself into a character; you must forge yourself into one.
Unknown

Watch your thoughts, for they become words. Watch your words, for they become actions. Watch your actions, for they become habits. Watch your habits, for they become character. Watch your character, for it becomes your destiny.

Unknown

Above all, we must realize that no arsenal, or no weapon in the arsenals of the world, is so formidable as the will and moral courage of free men and women.

Ronald Reagan

A hero is someone who over and over does the right thing even though no one is there to witness it.

Bob Kerrey

A man should be upright, not kept upright.

Emperor Marcus Aurelius

The success of my whole project is founded on the firmness of the conduct of the officer who will command it.
Frederick the Great

War must be carried on systematically, and to do it you must have men of character activated by principles of honor.
GEN George Washington

My character and good name are in my own keeping. Life with disgrace is dreadful. A glorious death is to be envied.
ADM Viscount Nelson

Strength of character does not consist solely in having powerful feelings, but in maintaining one's balance in spite of them.
MG Karl von Clausewitz

You may be whatever you resolve to be.
Stonewall Jackson

Private and public life are subject to the same rules; and truth and manliness are two qualities that will carry you through this world much better than policy, or tact, or expediency, or any other word that was ever devised to conceal or mystify a deviation from a straight line.
GEN Robert E. Lee

I am sorry, sir, that you are so little acquainted with m character as to suppose that my name is for sale at any price.
GEN Robert E. Lee

In war, character outweighs intellect.
GEN Werner von Fritsch

He must have character, which simply means that he knows what he wants and has the courage and determination to get it.
Field Marshall Viscount Wavell

Character is the bedrock on which the whole edifice of leadership rests.
GEN Matthew Ridgway

When faced with the challenge of events, the man of character has recourse to himself. His instinctive response is to leave his mark on action, to take responsibility for it, to make it his own business...
GEN Charles de Gaulle

Nearly all men can stand adversity, but if you want to test a man's character, give him power.
Abraham Lincoln

Leadership is a combination of character and strategy. If you must be without one, be without the strategy.
GEN H. Norman Schwarzkopf

One should not associate with people whose conduct is poor. In the Shih Chi it says If you don't know a man's character, investigate who his friends are.
Takeda Nobushige

True character always pierces through in moments of crisis...There are sleepers whose awakening is terrifying.
Napoleon Bonaparte

A military leader must possess as much character as intellect.
Napoleon Bonaparte

Strength of character does not consist solely in having powerful feelings, but in maintaining one's balance in spite of them.
MG Karl von Clausewitz

A sound body is good; a sound mind is better; but a strong and clean character is better than either.
Teddy Roosevelt

Some virtue is required which will provide the army with a new ideal, which, through the military elite, will unite the army's divergent tendencies and fructify its talent. This virtue is called character which will constitute the new ferment- character the virtue of hard times.
Charles de Gaulle

Of the many personal decisions that life puts upon the military officer, the main one is whether he chooses to swim upstream.
BG S.L.A. Marshall

It has been said that man's character is the reality of himself. I don't think a man's strength of character ever changes.
GEN Omar Bradley

A man of character in peace is a man of courage in war.
GEN Sir james Glover

Personal Courage

Be convinced that to be happy means to be free and that to be free means to be brave. Therefore do not take lightly the perils of war.
Thucydides

Part of the American dream is to live long and die young. Only those Americans who are willing to die for their country are fit to live.
General Douglas MacArthur

All men are timid on entering any fight. Whether it is the first or the last fight, all of us are timid. Cowards are those who let their timidity get the better of their manhood.
General George Patton Jr, War as I knew it 1947

Courage is fear holding on a minute longer.
General George Patton Jr

Fortune favors the brave.
Publius Terence

Out of every one hundred men, ten shouldn't even be there, eighty are just targets, nine are the real fighters, and we are lucky to have them, for they make the battle. Ah, but the one, one is a warrior, and he will bring the others back.
Heraclitus

Uncommon valor was a common virtue.
Admiral Chester Nimitz

Come on you sons of bitches! Do you want to live forever?
Gunnery Sergeant Dan Daly, 4 June 1918, Belleu Wood

Without a sign his sword the brave man draws, and asks no omen but his country's cause.
Homer, The Iliad

When Soldiers bave death, they drive him right into the enemy.'
Napoleon Bonaparte

I would rather have a Medal of Honor than be President of the United States.
Harry Truman

Courage above all things is the first quality of a Soldier.
MG Karl von Clausewitz

Life has a certain flavor for those who have fought and risked all that the sheltered and protected can never experience.
John Stuart Mill

Cowards die many times before their death; the valiant taste of death but once.
William Shakespeare

One man with courage makes a majority.
Andrew Jackson

The most drastic and usually the most effective remedy for fear is direct action.
William Burnham

It is better to suffer the worst right now than to live in perpetual fear of it.
Julius Caesar

Courage is being scared to death and saddled up anyway.
John Wayne

A hero is no braver than an ordinary person, but is braver five minutes longer.
Ralph Waldo Emerson

The men of the 5th Special Forces Group will continue to afford the enemy every opportunity to give their life for their country.
CSM Winston Clough

It were better to be a Soldier's widow then a coward's wife.
Thomas Aldrich

The courage of a Soldier is heightened by his knowledge of his profession
Vegetius 378

History does not long entrust the care of freedom to the weak or timid.
GEN Dwight Eisenhower

Courage is going from failure to failure without losing enthusiasm.
Prime Minister Winston Churchill

No man is worth his salt who is not ready at all times to risk his body, to risk his well being, to risk his life, in a great cause.
Teddy Roosevelt

The secret to happiness is freedom, and the secret to freedom, courage.
Thucydides

You have two choices in life, be wise or be brave, always choose to be brave.
Chinese Proverb

Valor is superior to numbers.
Vegetius

Sadness is mandatory; misery is optional.
Father of a Fallen Soldier

Courage is the finest of human qualities because it guarantees all the others.
Prime Minister Winston Churchill

Courage is like love: it must have hope for nourishment.
Napoleon Bonaparte

Courage! Do not fall back; in a little the place will be yours. Watch! When the wind blows my banner against the bulwark, you shall take it.
Joan of Arc

Courage may be taught as a child is taught to speak.
Euripides

Many would be cowards if they had courage enough.
Thomas Fuller

It is from numberless diverse acts of courage and belief that human history is shaped.
Robert Kennedy

The strongest, most generous, and proudest of all virtues is true courage.
Michel de Montagne

Untutored courage is useless in the face of educated bullets.
GEN George S. Patton

War is fear cloaked in courage.
GEN William Westmoreland

I am surrounded by fearful odds that will overcome me and my valiant men; but I am pleased to die fighting for my beloved country.
GEN Gregario Del Pilar

Bravery without forethought causes a man to fight blindly and desperately like a mad bull. Such an opponent must not be encountered with brute force, but may be lured into an ambush and slain.
Ts'ao Kung

Bravery does not of itself win battles.
H.H. Wilson

Freedom is the sure possession of those alone who have the courage to defend it.
Pericles

Being courageous does not mean never being scared; it means acting as you know you must even though you are undeniably afraid.
Archbishop Desmond Tutu

To fear is one thing. To let fear grabby you by the tail and swing you around is another.
Katherine Paterson

You can't be brave if you've only had wonderful things happen to you.
Mary Tyler Moore

Courage is nine-tenths context. What is courageous in one setting can be foolhardy in another and cowardly in a third.
Joseph Epstein

There is only one requirement for any of us, and that is to be courageous. Because courage, as you might know, defines all other human behavior. And I believe, because I've done a little of this myself—that pretending to be courageous is just as good as the real thing.
David Letterman, First broadcast after 9/11

Everything worth doing starts with being scared.
Art Garfunkel

Courage is the only magic worth having.
Erica Jong

Courage is not the absence of fear, but rather the judgment that something else is more important than fear.
Ambrose Redmoon

Only those who will risk going too far can possibly find out how far one can go.
T.S. Eliot

Never be afraid to try something new. Remember that amateurs built the Ark. Professionals built the Titanic.
Dave Barry

Do not fear death so much, but rather the inadequate life.
Bertolt Brecht

One crowded hour of glorious life is worth an age without a name.
Sir Walter Scott

My religious convictions teach me to feel safe in battle as if I were safe asleep in bed. The time of my death is fixed.
Stonewall Jackson

Theirs was not to reason why; Theirs was but to do, or die.
Tennyson, The Charge of the Light Brigade

Do not seek death. Death will find you. But seek the road which makes death a fulfillment.
Daj Hammarskjold

It is better to suffer the worst right now than to live in perpetual fear of it.
Julius Caesar

The Lacademonians ask not 'How many are the enemy?' but rather 'Where are the enemy?'
Athenian Saying

You gain strength, courage and confidence by every experience in which you must really stop to look fear in the face... You must do the thing you cannot do.
Eleanor Roosevelt

I wasn't a hero, but I served in a company of heroes.
Dick Winters, Band of Brothers

They are surely to be esteemed the bravest spirits who, having the clearest sense of both the pains and pleasures of life, do not on that account shrink from danger.
Thucydides

When Soldiers brave death, they drive him into the enemy's ranks.
Napoleon Bonaparte

I do not think that there is any man who would not rather be called brave than have any other virtue attributed to him.
Field Marshall Viscount Slim

The principle on which to manage an army is to set up one standard of courage which all must reach.
Sun Tzu

The more comfort the less courage there is.
Field Marshall Prince Aleksandr V

In sport, in courage, and in the sight of Heaven, all men meet on equal terms.
Prime Minister Winston Churchill

There is nothing like seeing the other fellow run to bring back your courage.
Field Marshall Viscount Slim

There are two kinds of courage, physical and moral, and he who would be a true leader must have both.
GEN Matthew Ridgway

All men are timid on entering any fight. Whether it is the first fight or the last fight, all of us are timid. Cowards are those who let their timidity get the better of their manhood.
GEN George S. Patton

The principle on which to manage an army is to set up one standard of courage which all must reach.
Sun Tzu

The history of free men is never really written by chance but by choice—their choice.
GEN Dwight Eisenhower

Whatever the dangers of the action we take, the dangers of inaction are far, far greater.
Tony Blair

A medal glitters, but it also casts a shadow.
Prime Minister Winston Churchill

The noble and courageous man is known by his patience in adversity.
The Inca Emperor Pachacutec

A brave captain is a root, out which as branches the courage of his Soldiers doth spring.
Sir Philip Sidney

As in all battles the dead and wounded came chiefly from the best and the bravest.
Field Marshall Lord Carver

You are uneasy; you never sailed with me before, I see.
Andrew Jackson

Without the threat of death there's no reason to live at all.
Brian Warner

The fear of war is worse than war itself.
Seneca

The universe is so vast and ageless that the life of one man can only be measured y the size of his sacrifice.
VA Rosewarne, RAF

The only thing we have to fear is fear itself.
Franklin D. Roosevelt

The finger of providence was upon me, and I escaped unhurt.
The Duke of Wellington

Whoever wants to see his own people again must remember to be a brave Soldier: that is the only way of doing it. Whoever wants to keep alive must aim at victory. It is the winners who do the killing and the losers who get killed.
Xenophon

Pay not attention to those who would keep you far from fire: you want to prove yourself a man of courage. If there are opportunities, expose yourself conspicuously. As for real danger, it is everywhere in war.
Napoleon Bonaparte

War is the realm of danger; therefore courage is the Soldier's first requirement.
MG Karl von Clausewitz

He was a brave man on such a day.
Spanish Proverb

A great many men, when they smell battle afar off, chafe to get into the fray.
GEN Ulysses S. Grant

In sport, in courage, and in the sight of Heaven, all men meet on equal terms.
Prime Minister Winston Churchill

There is no better ramrod for the back of a senior who is beginning to buckle than the sight of a junior who has kept his nerve.
BG S.L.A. Marshall

Leaders must study fear, understand it, and be prepared to cope with it.
GEN Alfred Gray

Not till I see day light ahead do I want to lead, but when danger threatens and others slink away I am and will be at my post.
GEN William T. Sherman

If you want to walk on water you have to get out of the boat.
John Ortberg

Fighting is like champagne. It goes to the heads of cowards as quickly as of heroes. Any fool can be brave on a battlefield when it's be brave or else be killed.
Margaret Mitchell

When men find they must inevitably perish, they willingly resolve to die with their comrades and with their arms in their hands.
Flavius Vegetius Renatus

In the course of battle, as long as you are able to advance on foot, never take a step backward.
Kai Ka'us Ibn Iskander

...for to die is to live forever in perpetual glory and honor.
Motecuhzoma, Aztec Emperor

There is one certain means by which I can be sure never to see my country's ruin: I will die in the last ditch.
William III

We beat them tonight or Molly Stark's a widow.
BG John Stark, Battle of Bennington, 1777

It so often happens that, when men are convinced that they have to die, a desire to bear themselves well and to leave life's stage with dignity conquers all other sensations.
Prime Minister Winston Churchill

Every position must be held to the last man: there must be no retirement. With our backs to the wall, and believing in the justice of our cause, each one of us must fight on to the end. The safety of our homes and the freedom of mankind alike depend on the conduct of each one of us at this critical moment.
Field Marshall Earl Haig

Patriotic

We shall defend our island whatever the cost may be; we shall fight on beaches, landing grounds, in fields, in streets and on the hills. We shall never surrender and even if, which I do not for the moment believe, this island or a large part of it were subjugated and starving, then our empire beyond the seas, armed and guarded by the British Fleet, will carry on the struggle until in God's good time the New World with all its power and might, sets forth to the liberation and rescue of the Old.

Prime Minister Winston Churchill (after the fall of France)

Don't one of you fire until you see the whites of their eyes. Powder is scarce and must not be wasted. Fire low! You are all marksmen and could kill a squirrel at a hundred yards. Reserve your fire and the enemy will all be destroyed

Israel Putnam, at the Battle of Bunker Hill and Breed's Hill, June 1775

Freedom itself was attacked this morning by a faceless coward. Freedom will be defended!

President George W. Bush, September 11, 2001

The best defense against terrorism is a strong offensive against terrorists. That work continues.
President George W. Bush, October 13, 2001

We will not tire, We will not falter, We will not fail.
President George W. Bush, October 26, 2001

For states that support terror, it is not enough that the consequences be costly-they must be devastating
President George W. Bush at a speech at The Citadel, Dec 11, 2001.

I gave them a fair warning
President George W. Bush

These acts shattered steel, but they cannot dent the steel of America's resolve.
President George W. Bush

What our enemies have begun, we will finish,
President George W. Bush on September 11, 2002

This will not be a campaign of half measures, and we will accept no outcome except victory.

President George W. Bush, March 20, 2003

The shepherd drives the wolf from the sheep's for which the sheep thanks the shepherd as his liberator, while the wolf denounces him for the same act as the destroyer of liberty. Plainly, the sheep and the wolf are not agreed upon a definition of liberty.

Abraham Lincoln

We shall meanly lose or nobly save the last hope of earth.

Abraham Lincoln

Peace will come soon to stay, and so come as to be worth keeping in all future time. It will then have proved that among free men there can be no successful appeal from the ballot to the bullet, and that they who take such appeal are sure their cases and pay the costs.

Abraham Lincoln

For a people who are free, and who mean to remain so, a well-organized and armed militia is their best security.
Thomas Jefferson: message to Congress, Nov. 1808

The price of freedom is eternal vigilance.
Thomas Jefferson

A young man who does not have what it takes to perform military service is not likely to have what it takes to make a living.
John F. Kennedy (JFK)

Ask not what your country can do for you; ask what you can do for your country.
John F. Kennedy (JFK)

Let every nation know, whether it wishes us well or ill, that we shall pay any price, bear any burden, meet any hardship, support any friend, oppose any foe, to assure the survival and success of liberty.
John F. Kennedy (JFK)

I can imagine no more rewarding a career. And any man who may be asked in this century what he did to make his life worthwhile, I think can respond with a good deal of pride and satisfaction: I served in the United States Navy

John F. Kennedy (JFK)

By profession I am a Soldier and take pride in that fact. But I am prouder -- infinitely prouder -- to be a father. A Soldier destroys in order to build; the father only builds, never destroys. The one has the potentiality of death; the other embodies creation and life. And while the hordes of death are mighty, the battalions of life are mightier still. It is my hope that my son, when I am gone, will remember me not from the battle field but in the home repeating with him our simple daily prayer, 'Our Father Who Art in Heaven.'

General Douglas MacArthur

It is foolish and wrong to mourn the men who died. Rather we should thank God that such men lived.

General George S. Patton, Jr

These are the times that try men's souls. The summer Soldier and the sunshine patriot will, in this crisis, shrink from the service of their country; but he that stands it now, deserves the love and thanks of man and woman. Tyranny, like hell, is not easily conquered; yet we have this consolation with us, that the harder the conflict, the more glorious the triumph. What we obtain too cheap, we esteem too lightly: it is dearness only that gives everything its value.
Thomas Paine, The American Crisis (1776)

The primary difference between Americans and everyone else is that Americans would rather die on their feet than live on their knees.
Unknown

Gunners will always fight together, drink together, laugh together, and mourn together.
Unknown

The eyes of the world are upon you. The hopes and prayers of liberty-loving people everywhere march with you.
GEN Dwight Eisenhower D-Day 1944

When you go home, tell them of us and say, for your tomorrow, we gave our today.
British 2ndDiv Memorial at Kohina

In the end, we will remember not the words of our enemies, but the silence of our friends.
Martin Luther King Jr.

Give us the tools and we will finish the job.
Prime Minister Winston Churchill

An Army lives in the shadow of its tradition. It looks to the heroic deeds of the past and the performance of its great Soldiers to enlighten and inspire its present membership.
GEN Fred Weyand

Better than honor and glory, and History's iron pen, Was the thought of duty done and the love of his fellow-men.
Richard Watson Gilder

And each man stands with his face in the light of his own drawn sword. Ready to do what a hero can.
Elizabeth Barrett Browning

In war, there are no unwounded Soldiers.
José Narosky

When our perils are past, shall our gratitude sleep?
George Canning

For what avail the plough or sail, or land or life, if freedom fail?
Ralph Waldo Emerson

We often take for granted the very things that most deserve our gratitude.
Cynthia Ozick

Let no vandalism of avarice or neglect, no ravages of time, testify to the present or to the coming generations, that we have forgotten, as a people, the cost of a free and undivided Republic.
John A. Logan

Where liberty dwells, there is my country.
Benjamin Franklin

He loves his country best who strives to make it best.
Robert G. Ingersoll

Then join hand in hand, brave Americans all! By uniting we stand, by dividing we fall.
John Dickinson

I can no other answer make, but, thanks, and thanks.
William Shakespeare

May the sun in his course visit no land more free, more happy, more lovely, than this our own country!
Daniel Webster

And they who for their country die shall fill an honored grave, for glory lights the Soldier's tomb, and beauty weeps the brave.
Joseph Drake

Who kept the faith and fought the fight; The glory theirs, the duty ours.
Wallace Bruce

This nation will remain the land of the free only so long as it is the home of the brave.
Elmer Davis

If our country is worth dying for in time of war let us resolve that it is truly worth living for in time of peace.
Hamilton Fish

Ours is the only country deliberately founded on a good idea.
John Gunther

If you are ashamed to stand by your colors, you had better seek another flag.
Unknown

Sometimes people call me an idealist. Well, that is the way I know I am an American. America is the only idealistic nation in the world.
Woodrow Wilson

My favorite thing about the United States? Lots of Americans, one America.
Val Saintsbury

Our great modern Republic. May those who seek the blessings of its institutions and the protection of its flag remember the obligations they impose.
Ulysses S. Grant

I think there is one higher office than president and I would call that patriot.
Gary Hart

Patriotism is easy to understand in America—it means looking out for yourself by looking out for your country.
Calvin Coolidge

We dare not forget that we are the heirs of that first revolution.
John F. Kennedy

If I were an American, as I am an Englishman, while a foreign troop was landed in my country I never would lay down my arms, never! Never! Never!
William Pitt

Let us show the whole world that a Freeman, contending for liberty on his own ground, is superior to any slavish mercenary on Earth.
George Washington

Through their deeds, the dead of battle have spoken more eloquently for themselves than any of the living ever could. But we can only honor them by rededicating ourselves to the cause for which they gave a last full measure of devotion.
Abraham Lincoln

Give me liberty or give me death.
Patrick Henry

I will not disgrace the Soldier's arms, nor abandon the comrade who stands at my side; ut whether alone or with many, I will fight to defend things sacred and profane. I will hand down my country not lessened, but larger and better than I have received it.
Athenian Oath

God grant liberty only to those who love it and are always ready to guard and defend it.
Daniel Webster

Remember the Alamo.
Battle Cry at San Jacinto

Go tell the Spartans, stranger passing by. Obedient to their laws we lie.
Simonides of Keos, inscription at Thermopylae

Here once the embattled farmers stood, and fired the shot heard round the world.
Ralph Waldo Emerson

We hold these truths to be self-evident that all men are created equal; that they are endowed by their Creator with certain unalienable rights; that among these are life, liberty, and the pursuit of happiness.
Thomas Jefferson from the Declaration of Independence

I shall return.
GEN Douglas Macarthur

Give me your tired, your poor, your huddled masses yearning to breathe free.
Emma Lazarus

Give the American people a good cause, and there's nothing they can't lick.
John Wayne

I regret that I have but one life to give for my country.
Nathan Hale

This empire has been acquired y men who knew their duty and had the courage to do it, who in the hour of conflict had the fear of dishonor always present to them, and who, if ever they failed in an enterprise, would not allow their virtues to be lost to their country, but freely gave their lives to her as the fairest offering which they could present at her feast.
Pericles

Our country: in her intercourse with foreign nations may she always be in the right; bout our country, right or wrong!
ADM Stephen Decatur

Certainly no man has more that should make life dear to him than I have, in the affection of my home; but I do not want to survive the independence of my country.
Stonewall Jackson

The man who is willing to fight for his country is finally the full custodian of its security.
BG S.L.A. Marshall

I admire men who stand up for their country in defeat, even though I am on the other side.
Prime Minister Winston Churchill

Sending Americans into battle is the most profound decision a president can make. The technologies of war have changed. The risks and suffering have not. For the brave Americans who bear the risk, no victory is free from sorrow.
President George W. Bush

When we assumed the Soldier, we did not lay aside the Citizen.
George Washington

The atom bomb was no 'great decision.'...It was merely another powerful weapon in the arsenal of righteousness.
Harry Truman

War is the remedy that our enemies have chosen, and I say let us give them all they want.
GEN William T. Sherman

Stand firm ye boys from Maine.
COL Joshua Chamberlain

Ladies and gentlemen, we got him.
Paul Bremer on the capture of Saddam Hussein

Tell the men to fire faster and not to give up the ship. Fight her til she sinks.
CPT James Lawrence, 1813, order as he lay dying during the battle between the USS Chesapeake and HMS Shannon

Fight for your country- that is the best, the only omen!
Homer, The Iliad

Discipline

I believe that no other new regiment will ever have the discipline we have now; we all work.

COL Joshua Chamberlain, Commander 20th Maine

It is absurd to believe that Soldiers who cannot be made to wear the proper uniform can be induced to move forward in battle. Officers who fail to perform their duty by correcting small violations and in enforcing proper conduct are incapable of leading.

General George S. Patton Jr., April 1943

You cannot be disciplined in great things and indiscipline in small things. Brave un-disciplined men have no chance against the discipline and valor of other men. Have you ever seen a few policemen handle a crowd?

General George S. Patton Jr, May 1941

If you can't get them to salute when they should salute and wear the clothes you tell them to wear, how are you going to get them to die for their country?
General George Patton Jr

The man who can't make a mistake can't make anything.
Abraham Lincoln

You are always on parade...There is no such thing as 'a good field Soldier.' You are either a good Soldier or a bad Soldier.
GEN George Patton

The noncommissioned officer wearing the chevron is supposed to be the best Soldier in the platoon and he is supposed to know how to perform all the duties expected of The American Soldier.
GEN Omar Bradley

He who lives without discipline dies without honor.
Icelandic Proverb

Discipline is the soul of the Army. It makes small numbers formidable; procures success to the weak and esteem to all.
George Washington

Military organizations and success in battle depend upon discipline and a high sense of honor.
GEN Omar Bradley

You owe it to your men to require standards which are for their benefit even though they may not be popular at the moment.
GEN Bruce Clarke

Our troops are capable of the best discipline. If they lack it, leadership is faulty.
GEN Dwight Eisenhower

The discipline which makes the Soldiers of a free country reliable in battle is not to be gained by harsh or tyrannical treatment. On the contrary, such treatment is far more likely to destroy than to make an army. It is possible to impart instruction and to give commands in such a manner and such a tone of voice to inspire in the Soldier no feeling but an intense desire to obey, while the opposite manner and tone of voice cannot fail to excite strong resentment and a desire to disobey. The one mode or the other of dealing with subordinates springs from a corresponding spirit in the breast of the commander. He who feels the respect which is due to others cannot fail to inspire in them regard for himself, while he who feels, and hence manifests, disrespect toward others, especially his inferiors, cannot fail to inspire hatred against himself.

Major General John M. Schofield

Maneuvering with an army is advantageous; with an undisciplined multitude, most dangerous.

Sun Tzu

Discipline is simply the art of making the Soldiers fear their officers more than the enemy.

Helvetius

It was a magnificent display of trained and disciplined valour, and its assault failed of success only because dead men can advance no further.
Battle of the Somme

Praise in public but discipline in private.
Unknown

The great end of education is to discipline rather than to furnish the mind; to train it to the use of its own powers, rather than fill it with the accumulation of others.
Tryon Edwards

Bravery comes along as a gradual accumulation of discipline.
Buzz Aldrin

My troops are good and well disciplined...
Frederick the Great

Be men, my friends! Discipline fill your hearts!
Homer, The Illiad

For where there is no one in control nothing useful or distinguished can be done.
Xenophon

Discipline is the soul of an army. It makes small numbers formidable; procures success to the weak, and esteem to all.
George Washington

The more modern war becomes, the more essential appear the basic qualities that from the beginning of history have distinguished armies from mobs. The first of these is discipline.
Field Marshall Viscount Slim

You pick out the big men! I'll make them brave.
Pyrrhus

After the organization of troops, military discipline is the first matter that presents itself.
Field Marshall Maurice Comte de Saxe

My son, put a little order into your corps, it wants it badly.
Napoleon Bonaparte

Impress upon your officers that discipline cannot be attained without constant watchfulness on their part.
GEN Robert E. Lee

Discipline is not made to order; cannot be created off-hand...
COL Charles Ardant du Picq

Discipline can only be obtained when all officers are so imbued with the sense of their awful obligation to their men and to their country that they cannot tolerate negligence.
GEN George S. Patton

Popularity, however desirable it may be to individuals, will not form, or feed, or pay an army; will not enable it to march, and fight; will not keep it in a state of efficiency for long and arduous service.
The Duke of Wellington

It is the discipline which is the soul of armies, as indeed the soul of power in all intelligence.
MG Joshua Chamberlain

An army is effective in proportion to its discipline.
MG Joshua Chamberlain

Discipline is the basis of military efficiency.
MG John Lejeune

The idea that real discipline is best instilled by fear of punishment is a delusion- especially in these days of open-order fighting.
Sir Basil Liddell Hart

Within our system, that discipline is nearest perfect which assures to the individual the greatest freedom of thought and action while at all times promoting his feeling of responsibility toward the group.
BG S.L.A. Marshall

One of the primary purposes of discipline is to produce alertness. A man who is so lethargic that he fails to salute will fall an easy victim to the enemy.
GEN George S. Patton

The strength of an army lies in strict discipline and undeviating obedience to its officers.
Thucydides

Be men, my friends! Discipline fill your hearts! Dread what comrades say of you here in bloody combat!
Homer, The Iliad

All military history ... records the triumphs of discipline and courage far more frequently than numbers and resources.
GEN Robert E. Lee

The importance and utility of thorough discipline should be impressed on officers and men on all occasions by illustrations taken from the experience of the instructor...
GEN Robert E. Lee

In all armies, obedience of the subordinates to their superiors must be exacted...
Mao Tse-tung

The firmness requisite for the real business of fighting is only to be obtained by a constant course of discipline and service.
George Washington

Discipline, to which officer and private alike are subjected, was, in my opinion, the only basis on which an army could be effectively trained for war.
GEN Erich Ludendorff

All human beings have an innate resistance to obedience.
GEN George S. Patton

No sane man is unafraid in battle, but discipline produces in him a form of vicarious courage which, with his manhood, makes for victory.
GEN George S. Patton

Discipline seeks through drill to instill into all ranks this sense of unity, by requiring them to obey orders as one man.
GEN Viscount Alexander of Tunis

Loyalty

I knew wherever I was that you thought of me and if I got into a tight place you would come, if alive.

William T. Sherman

In all military works it is written: to train samurai to be loyal separate them when young, or treat them according to their character.

Tokugawa Ieyasu

Loyalty is the marrow of honor.

Field Marshall Paul von Hindenburg

Those who are naturally loyal say little about it, and are ready to assume it in others.

Sir Basil Liddell Hart

An officer should make it a cardinal principle of life that by no act of commission or omission on his part will he permit his immediate superior to make a mistake.
GEN Malin Craig

We learn from history that those who are disloyal to their own superiors are most prone to preach loyalty to their subordinates.
Sir Basil Liddell Hart

There is a great deal of talk about loyalty from the bottom to the top. Loyalty from the top down is even more necessary and much less prevalent.
GEN George S. Patton Jr.

The very touchstone of loyalty is that just demands will be put upon it.
S.L.A. Marshall

I also learned that the discipline demanded from the Soldier must become loyalty in the officer.
Field Marshall Viscount Montgomery

The essence of loyalty is the courage to propose the unpopular, coupled with a determination to obey, no matter how distasteful the ultimate decision.
LTG Victor Krulak

When we are debating an issue, loyalty means giving me your honest opinion, whether you think I'll like it or not.'
GEN Colin Powell

...the basic principles of the military services are unchangeable. Courage and candor, obedience and comradeship, love of fatherland and loyalty to the State: these are ever the distinguishing characteristics of the Soldier and sailor. Building character through intelligent training and education is always the first and greatest goal.
Grand ADM Erich Raeder

Stand with anybody that stands right, stand with him while he is right and part with him when he goes wrong.
Abraham Lincoln

Creating a loyal following among your team means that the leader shows loyalty to the crew.
H. Ross Perot

Let us set up a standard around which the brave and loyal can rally.
Prime Minister Winston Churchill

Those who would be military leaders must have loyal hearts, eyes and ears, claws and fangs.
Zhuge Liang

Without people loyal to them, they are like someone walking at night, not knowing where to step.
Zhuge Liang

Loyalty is the big thing, the greatest battle asset of all.
BG S.L.A. Marshall

I also learnt that the discipline demanded from the Soldier must become loyalty in the officer.
Field Marshall Viscount Montgomery

The essence of loyalty is the courage to propose the unpopular, coupled with a determination to obey, no matter how distasteful the ultimate decision. And the essence of leadership is the ability to inspire such behavior.
LTG Victor Krulak

Tactics

In tactics every engagement, great or small, is a defensive one if we leave the initiative to the enemy.

MG Carl von Clausewitz

Tactics is an art based on the knowledge of how to make men fight with maximum energy against fear, a maximum which organization alone can give.

COL Charles du Picq

I do not believe that the officers of the regiment ever discovered that I had never studied the tactics that I used.

Ulysses S. Grant

Nine-tenths of tactics are certain, and taught in books; but the irrational tenth is like kingfisher flashing across the pool and that is the test of generals.

T.E. Lawrence

Tactics are the cutting edge of strategy, the edge which chisels out the plan into an action; consequently the sharper this edge is the clearer will be the result.
MG J.F.C. Fuller

There is only one tactical principle which is not subject to change. It is: to use the means at hand to inflict the maximum amount of wounds, death and destruction on the enemy in the minimum time.
George S. Patton Jr.

All tactics since the earliest days have been based on evaluating an equation in which x=mobility, y=armor, and z=hitting power.
Viscount Wavell

War is nothing but a duel on an extensive scale.
MG Karl von Clausewitz

All men can see these tactics whereby I conquer, but what none can see is the strategy out of which victory evolved.
Sun Tzu

Sir, my strategy is one against ten, my tactics are ten against one.
The Duke of Wellington

A tactical success is only really decisive, if it is gained at the strategically correct spot.
Field Marshall Helmuth von Moltke

Contact, a word which perhaps better than any other indicates the dividing line between strategy and tactics.
Admiral Alfred Thayer Mahan

In peace we concentrate so much on tactics that we are apt to forget that it is merely the handmaiden of strategy.
Sir Basil Hart

So far as I can see strategy is eternal, and the same and true: but tactics are the ever changing languages through which it speaks.
T.E. Lawrence

I rate the skillful tactician above the skillful strategist, especially him who plays the bad cards well.
Viscount Wavell

Any blow, to be successful, must be sudden and hard.
Robert E. Lee

I was too weak to defend, so I attacked.
Robert E. Lee

Always employ outposts. Always utilize patrols. Always keep a reserve.
Erwin Rommel

Things may come to those who wait, but only those things left behind by those who hustle.
Abraham Lincoln

Never use intuition. It will never substitute for good intelligence.
GEN Omar Bradley

Get there first with the most men.
Nathan Bedford Forrest

Be there the firstest, with the mostest.
Ulysses S. Grant

The second assault on the same problem should come from a totally different direction.
Tom Hirschfield

Charging beats retreating.
Malcolm Forbes

In tactics, action is the governing rule of war.
Ferdinand Foch

The purpose of defense is to break the strength of the attacker, bleed him white—then attack.
Von Leeb

Surprise is the key to victory.
Erfurth

NCOs must love initiative and must hold what ground they have gained to the utmost. It often happens that a sergeant or even a corporal may decide a battle by the boldness with which he seizes a bit of ground and holds it.
GEN Blackjack Pershing

Don't hit at all if it is honorably possible to avoid hitting; but never hit soft!
Teddy Roosevelt

In war, there is no substitute for victory.
GEN Douglas MacArthur

Hit hard, hit fast, hit often.
ADM William Halsey Jr.

In every battle there comes a time when both sides consider themselves beaten: then he who continues to attack, wins.
Ulysses S. Grant

A battle sometimes decides everything; and sometimes the most trifling thing decides the fate of a battle.
Napoleon Bonaparte

A good plan violently executed right now is better than 100 perfect plans put off until next week.
GEN Douglas MacArthur

Take time to deliberate…but when the time for action has arrived, stop thinking and go in.
Napoleon Bonaparte

I may lose a battle, but I will lose no time.
Napoleon Bonaparte

The proof of battle is action, proof of words, debate. No time for speeches now, it's time to fight.
Homer, The Iliad

In tactics, action is the governing rule of war.
Ferdinand Foch

A big butcher's bill is not necessarily evidence of good tactics.
Field Marshall Viscount Wavell

The difficulty of tactical maneuvering consists in turning the devious into the direct, and misfortune into gain.
Sun Tzu

So in war, the way is to avoid what is strong and to strike at what is weak.
Sun Tzu

He who can modify his tactics in relation to his opponent and thereby succeed in winning, may be called a heaven-born captain.
Sun Tzu

Military tactics are like unto water; for water in its natural course runs away from high places and hastens downwards...
Sun Tzu

In making tactical dispositions, the highest pitch you can attain is to conceal them.
Sun Tzu

If there is one thing you can count on in war it is that there is nothing you can count on in war.
Richard M. Watt

In the absence of orders, go find something and kill it.
Erwin Rommell

It is not simply the weapons one has in one's arsenal that give one flexibility, but the willingness and ability to use them.
Mao Tse-tung

The sharp general takes into account not only probably dangers, but also those which may be totally unexpected.
The Emperor Maurice

The objective is not the occupation of a geographical position but the destruction of the enemy force.
GEN Piotr Rumyantsev

Integrity

Commanders must have integrity; without integrity, they have no power. If they have no power, they cannot bring out the best in their armies. Therefore, integrity is the hand of warriorship.

Sun Bin

If you choose Godly, honest men to be captains of horse, honest men will follow them.

Oliver Cromwell

In order to be a leader, a man must have followers. And to have followers, a man must have their confidence. Hence the supreme quality for a leader is unquestionable integrity.

GEN Dwight Eisenhower

Integrity, of course, embraces much more than just simple honesty. It means being true to your men, true to your outfit, and above all true to yourself.

LTG Sir James Glover

Integrity is one of those words that many people keep in that desk drawer labeled 'too hard.'
ADM James Stockdale

Personal honor is the one thing valued (by a good man) more than life itself.
C. Markos

One needs to be slow to form convictions; but once formed they must be defended against the heaviest odds.
Mahatmas Gandhi

What you want to be eventually, that you must be every day; and by and by the quality of your deeds will get down to your soul.
Frank Crane

Associate with men of good quality, if you esteem your own reputation; for it is better to be alone than in bad company.
George Washington

It makes a difference to all eternity whether we do right or wrong today.
James Clark

The world has achieved brilliance without conscience. Ours is a world of nuclear giants and ethical infants.
GEN Omar Bradley

The man who speaks the truth is always at ease.
Persian proverb

Always do right. This will surprise some people, and astonish the rest.
Mark Twain

Someone will always be looking at you as an example of how to behave. Don't let that person down.
H. Jackson Brown Jr.

We must adjust to changing times and still hold to unchanging principles.
Jimmy Carter

Be sure you are right, then go ahead.
Davy Crockett

The time is always right to do what is right.
Martin Luther King Jr.

The hottest places in hell are reserved for those who in times of moral crisis preserve their neutrality.
Robert F. Kennedy

The ordinary Soldier has a surprisingly good nose for what is true and what is false.
Erwin Rommel

I never had a policy; I have just tried to do my very best each and every day.
Abraham Lincoln

If you have integrity, nothing else matters. If you don't have integrity, nothing else matters.
Alan Simpson

The right to do something does not mean that doing it is right.
William Safire

To thine own self be true, and it must follow, as the night the day, thou canst not then be false to any man.
William Shakespeare, Hamlet

A man has to live with himself, and he should see to it that he always has good company.
Charles Evans Hughes

You do not wake up one morning a bad person. It happens by a thousand tiny surrenders of self-respect to self-interest.
Robert Brault

My goal in life is to be as good of a person my dog already thinks I am.
Unknown

To know what is right and not do it is the worst cowardice.
Confucius

Just do good, don't worry about the road ahead.
Monk Wansong

I hope I shall always possess firmness and virtue enough to maintain what I consider the most enviable of all titles, the character of 'an honest man.'
George Washington

An honest man is the noblest work of God.
Alexander Pope

Honesty is the best policy. It's the policy.
Geraud Darnis, President Carrier Corporation

The real honest man is honest from conviction of what is right, not from policy.
GEN Robert E. Lee

Selfless Service

A candle loses nothing by lighting another candle.

Unknown

What a great difference there is between giving advice and lending a hand.

Robert Cloward

The noblest service comes from nameless hands, and the best servant does his work unseen.

Unknown

It is high time the ideal of success should be replaced with the ideal of service.

Albert Einstein

When one helps another, both are strong.

German Proverb

Service makes men competent.
Lyman Abbott

We make a living by what we get, but we make a life by what we give.
Prime Minister Winston Churchill

The man who lives for himself is a failure; the man who lives for others is a true success.
Norman Peale

I don't know what your destiny will be, but one thing I know: the only ones among you who will be really happy are those who will have sought and found how to serve.
Dr. Albert Schweitzer

The obligation I am under to my countrymen for the great honor they have conferred on me by returning me to the highest office within their gift, and the further obligation resting on me to render to them the best services within my power.
Ulysses Grant

He who makes his chief business seeking happiness never finds it. If he makes his chief business service to others, happiness will seek him
Dr. Eric Russ

Do all the good you can, by all the means you can, in all the ways you can, in all the places you can, to all the people you can, as long as ever you can.
John Wesley

A stingy man is always poor.
French Proverb

Do not wait for extraordinary circumstances to do good actions; try to use ordinary situations.
Richter

When a friend is in trouble, don't annoy him by asking if there is anything you can do. Think up something appropriate, and do it.
E.W. Howe

It is one of the most beautiful compensations of life that no man can sincerely try to help another without helping himself.
John P. Webster

One of the most durable satisfactions in life is to lose one's self in one's work.
Henry Fosdick

Teamwork

Each of us is the company.

William Hewlett

All for one and one for all.

Alexandre Dumas The Younger

Coming together is a beginning; keeping together is progress; working together is success.

Henry Ford Sr.

No member of a crew is praised for his rugged individuality in rowing.

Ralph Waldo Emerson

Light is the task where many share the toil.

Homer

Snowflakes are one of nature's most fragile things; but look what they can do when they stick together.
Unknown

A major reason capable people fail to advance is that they don't work well with their colleagues.
Lee Iacocca

The plan of one man may be faulty, that of two will be better.
Chuang Tse

Any idea can turn into dust or magic, depending on the talent that rubs against it.
William Bernbach

No man is wise enough by himself.
Plautus

Where there is unity there is always victory.'
Plubilius Syrus

Cooperation is spelled in two letters; WE.
George Verity

There's just three things I ever say to my team. If anything goes bad, then I did it. If anything goes semi-good, then we did it. If anything goes real good, then you did it.
Paul Bear Bryant

I not only use all the brains I have, but all I can borrow.
Woodrow Wilson

We must, indeed, all hang together, or, most assuredly, we shall all hang separately.
Benjamin Franklin

It is the willingness of people to give of themselves over and above the demands of the job that distinguishes the great from the merely adequate organizations.
Peter Drucker

Alone we can do so little; together we can do so much.
Helen Keller

Coming together is a beginning. Keeping together is progress. Working together is success.
Henry Ford

Talent wins games, but teamwork and intelligence wins championships.
Michael Jordan

Teamwork is the ability to work towards a common vision, the ability to direct individual accomplishment toward organizational objectives. It is the fuel that allows common people to obtain uncommon results.
Unknown

When one helps another both are strong.
German Proverb

Any idea can turn into dust or magic, depending on the talent that rubs against it.
William Bernbach

There are parts of a ship which taken by themselves would sink. The engine would sink. The propeller would sink. But when the parts of a ship are built together, they float.
Ralph Waldo Emerson

Every great man is always being helped by everybody; for his gift is to get good out of all things and persons.
John Ruskin

Its not going to get fixed until everybody takes responsibility for it. Like one of those old West Virginia proverbs I always talk about. A man benefits directly from a mistake, relative to how much it bothers him. I can tell you I didn't sleep very much so it bothers the hell out of me.
Nick Saban

As a business, we will either make dust, or eat it.
US West Communications Motto

A group becomes a team when each member is sure enough of himself and his contribution to praise the skills of the others.
Norman Shidle

Teamwork divides the task and multiplies the success.
Unknown

No one can whistle a symphony. It takes a whole orchestra to play it.
H.E. Luccock

Individual commitment to a group effort - that is what makes a team work, a company work, a society work, a civilization work.
Vince Lombardi

The nice thing about teamwork is you always have others on your side.
Margaret Carty

Cooperation is the thorough conviction that nobody can get there unless everybody gets there.
Virginia Burden

Sticks in a bundle are unbreakable.
Kenyan Proverb

Contrary to popular belief, there most certainly is an 'I' in 'team.' It is the same 'I' that appears three times in 'responsibility.'
Amber Harding

Never doubt that a small group of thoughtful, committed people can change the world. Indeed, it is the only thing that ever has.
Margaret Meade

The way a team plays as a whole determines its success. You may have the greatest bunch of individual stars in the world, but if they don't play together, the club won't be worth a dime.
Babe Ruth

I am a member of a team, and I rely on the team, I defer to it and sacrifice for it, because the team, not the individual, is the ultimate champion.
Mia Hamm

In union there is strength.
Aesop

Remember upon the conduct of each depends the fate of all.
Alexander the Great

Getting good players is easy. Getting them to play together is the hard part.
Casey Stengel

Winning is great, but sharing the victory with teammates is just as special.
There is no 'I' in 'team.'
Quarterback Tom Watson

We are apt to forget that we are only one of a team; that in unity there is strength and that we are strong only as long as each unit in our organization functions with precision.
Samuel Tilden

Two men in a burning house must not stop to argue.
Ashanti Proverb

Try your best to get to the top if that's where you want to go. But know that the more people you try to take along with you, the faster you'll get there and the longer you'll stay there.
James Autry

Although a single twig will break, a bundle of twigs is strong.
Tucumsia

An army is a team. It eats, sleeps, lives and fights as a team. All this stuff you've been hearing about individuality is a bunch of crap.
GEN George S. Patton

Four brave men who do not know each other will not dare attack a lion. Four less brave men, but knowing each other well, sure of their reliability and consequently of their mutual aid, will attack resolutely.
COL Charles Ardent du Picq

Morale & Esprit de Corps

Recognition for a job well done is high on the list of motivating influences for all people; more important, in many instances, than compensation itself.

John Wilson

The number one motivator of people is feedback on results.

Felix Doc Blanchard

Victory is spirit.

Anatole France

You can employ men and hire hands to work for you, but you must win their hearts to have them work with you.

Tiorio

You can have all the material in the world, but without morale it is largely ineffective.

GEN George C. Marshall

It is morale that wins the victory. With it, all things are possible; without it, everything else—planning, preparation and production, count for nothing.
GEN George C. Marshall

The best Soldiers show no rashness. The best fighters display no anger. The best conqueror seeks no revenge.
Taoism

A battle is lost less through the loss of men than by discouragement.
Frederick The Great

A battle is won by those who are firmly resolved to win it.
Tolstoy

You cannot expect a Soldier to be a proud Soldier if you humiliate him. You cannot expect him to be brave if you abuse and cower him. You cannot expect him to be strong if you break him. You cannot expect him to fight and die for our cause if your Soldier has not been treated with the respect and dignity which fosters unit esprit and personal pride.
GEN Delos Emmons

A good leader inspires men to have confidence in him; a great leader inspires men to have confidence in themselves.
Unknown

You can buy a man's time, you can buy his physical presence in a given place; you can even buy a measured number of his skilled muscular motions per hour. But you cannot buy enthusiasm...you cannot buy loyalty... you cannot buy the devotion of hearts, minds, or souls. You must earn these.
Charles Frances

Confidence is contagious. So is lack of confidence.
Michael O'Brien

The sinews of war are five- men, money, materials, maintenance and morale.
Ernest Hemmingway

When a general complains about the morale of his troops, the time has come to look at his own.
GEN George C. Marshall

The morale of the Soldier is the greatest single factor in war.
Field Marshall Bernard Montgomery

Of all the people I have met in this world, the Marines have the cleanest bodies, the filthiest minds, the highest morale, and the lowest morals of all. Thank God for the Marines.
Eleanor Roosevelt

You are all aware that it is not numbers or strength that bring the victories in war. No, it is when one side goes against the enemy with the gods' gift of a stronger morale that their adversaries, as a rule, cannot withstand them.
Xenophon

In war everything depends on morale; and morale and public opinion comprise the better part of reality.
Napoleon Bonaparte

Morale makes up three quarters of the game; the relative balance of man-power accounts only for the remaining quarter.
Napoleon Bonaparte

One fights well when the heart is light.
Napoleon Bonaparte

The unfailing formula for production of morale is patriotism, self-respect, discipline, and self-confidence within a military unit, joined with fair treatment and merited appreciation from without.
GEN Douglas MacArthur

Morale, only morale, individual morale as a foundation under training and discipline, will bring victory.
Field Marshall Viscount Slim

Morale is a state of mind. It is steadfastness and courage and hope.
GEN George C. Marshall

Machines are nothing without men. Men are nothing without morale.
ADM Ernest King

Morale is the big thing in war.
Field Marshall Viscount Montgomery

Do not place military cemeteries where they can be seen by replacements marching to the front. This has a very bad effect on morale, even if it adds to the pride of the Graves Registration Service.
GEN George S. Patton

Very many factors go into the building up of sound morale in an army, but one of the greatest is that men be fully employed at useful and interesting work.
Prime Minister Winston Churchill

The morale of the Soldier is the greatest single factor in war and the best way to achieve a high morale in war-time is by success in battle.
Field Marshall Viscount Montgomery

Morale is a state of mind. It is that intangible force which will move a whole group of men to give their last ounce to achieve something, without counting the cost to themselves; that makes them feel they are part of something greater than themselves.
Field Marshall Viscount Slim

Airborne esprit is further enhanced by the Soldier's knowledge that he repeatedly does something many men cannot force themselves to do; jump out of airplanes...
GEN Hamilton Howze

Adapt yourself to the environment in which your lot has been cast, and show true love to the fellow-mortals with whom destiny has surrounded you.
Emperor Marcus Aurelius

After my death will you find a king who deserves such men?
Alexander the Great

The worst cowards, banded together, have their power.
Homer, the Iliad

Victory and disaster establish indestructible bonds between armies and their commanders.
Napoleon Bonaparte

Solidarity and confidence cannot be improvised...It is time we should understand the lack of power in mob armies.
COL Charles Ardant du Picq

A wise organization ensures that the personnel of combat groups changes as little as possible, so that comrades in peacetime shall be comrades in war.
COL Charles Ardant du Picq

The thing in any organization is the creation of a soul.
GEN George S. Patton

The best tactical results obtain from those dispositions and methods which link the power of one man to that of another.
BG S.L.A. Marshall

My first wish would be that my military family, and the whole army, should consider themselves as a band of brothers, willing and ready to die for each other.
George Washington

A mysterious fraternity born out of smoke and danger of death.
Stephen Crane, The Red badge of Courage

Great achievements in war and peace can only result if officers and men form an indissoluble band of brothers.
Field Marshall Paul von Hindenburg

I hold it to be one of the simplest truths of war that the thing which enables an infantry Soldier to keep going with his weapons is the near presence or the presumed presence of a comrade.
BG S.L.A. Marshall

Down south his men were on patrol, or digging new perimeters, or dying, and he was nothing if he did not share that misery.
James Webb

One of the lasting truths about being a Soldier is that friendships formed with comrades in arms are the deepest and most enduring.
GEN Fred Franks

I never saw a more confident army. The Soldiers think I know everything and that they can do anything.
GEN William T. Sherman

...esprit de corps means love for one's military legion...
LTG Chesty Puller

I wanted to imbue my crews with enthusiasm and a complete faith in their arm and to instill in them a spirit of selfless readiness to serve in it.
ADM Karl Donitz

One's own division is necessarily number one.
GEN Maxwell Taylor

Special Operations

The men with painted faces. They are like the cobra. They strike a deadly blow, then are gone leaving the dead for their ancestors.

From the novel Charlie Mike

There's still a lot of bad guys out there, guys that need killing.

SF NCO in Al Hillah, Iraq

For the operators, whom a wise commander uses with great skill and forethought, and whom the fool throws away in ignorance and contempt.

Greg Walker

We want to be in a situation under maximum pressure, maximum intensity, and maximum danger. When it's shared with others, it provides a bond which is stronger than any tie that can exist.

Seal Team Six Officer

> ...were our eyes and ears on the [battlefield]... Special Forces were the glue that held the coalition together.
>
> GEN Norman Schwarzkopf

> The only easy day was yesterday.
>
> Navy SEAL motto

> Somewhere a True Believer is training to kill you. He is training with minimal food or water, in austere conditions, training day and night. The only thing clean on him is his weapon and he made his web gear. He doesn't worry about what workout to do - his ruck weighs what it weighs, his runs end when the enemy stops chasing him. This True Believer is not concerned about 'how hard it is;' he knows either he wins or dies. He doesn't go home at 17:00, he is home. He knows only The Cause. Still want to quit?
>
> SFAS Cadre Member

My dad was in Special Forces, so he was tough.
Tiger Woods

Humans are more important than hardware. Quality is better than quantity. SOF cannot be mass produced. Competent SOF cannot be created after emergencies occur. Most special operations require non-SOF assistance.
SOF Truths

I'd rather go down the river with seven studs than with a hundred shitheads.
COL Charlie Beckwith

When they say 'special operations', they are not kidding: these people are special. From an operational standpoint, physiological standpoint, they can do things that nobody else can do.
Dennis Grahn

Think like a bank robber.
GEN Wayne Downing

> You better get right with God before you do this Rangering thing
> CSM Donald Purdy

> I took a different route from most and came into Special Forces.
> COL Nick Rowe

> I've always been an aficionado of grenades... pistol fascination never gripped me.
> SGM Billy Waugh

> You know what I think. It doesn't really matter what I think. Once that first bullet goes past your head, politics and all that other crap goes right out the window.
> Hoot from Blackhawk Down

> A lot of folks want to wear the beret, but only a few want to carry the rucksack.
> COL AJ 'Bo' Baker

Good morning, gentlemen. You are now POWs.
Dick Meadows

You are to let nothing interfere with this operation. Our mission is to rescue prisoners, not to take prisoners. If there's been a leak, we'll know it as soon as the second or third chopper sets down... We'll make them pay for every foot.
COL Arthur Bull Simons

The Green Beret isn't just a piece of headgear; it is a symbol of all that is good and right about America. It represents the finest Soldiers ever to take the battlefield.
ADM William McRaven

The Green Beret is a symbol of excellence, a badge of courage, a mark of distinction in the fight for freedom.
John F. Kennedy

I am sure that the Green Beret will be a mark of distinction in the trying times ahead.
John F. Kennedy

I asked for a few Americans. They brought with them the courage of a whole army.
GEN Dostum on the Green Berets, Nov 2001

There is another type of warfare- new in its intensity, ancient in its origin- war by guerillas, subversives, insurgents, assassins; war by ambush instead of by combat, by infiltration instead of aggression, seeking victory by eroding and exhausting the enemy instead of engaging him.
John F. Kennedy

Ranger instructors should be like dogs. Eat once a day and run twice a day.
CSM Herbert Kirkover

These are the boys of Pointe du Hoc. These are the men who took the cliffs. These are the champions who helped free a continent. These are the heroes who helped end a war.
Ronald Reagan

The military value of a partisan's work is not measured by the amount of property destroyed or the number of men killed or captured, but by the number he keeps watching.
COL John S. Mosby

Guerilla war is far more intelligent than a bayonet charge.
T.E. Lawrence

Advance like foxes, fight like lions, and fly like birds.
Northeastern American Indian Tactical Maxim

I selected 'Rangers' because few words have a more glamorous connotation in American military history... men who exemplified such high standards of individual courage, initiative, determination, and ruggedness.
BG Lucian Truscott on the term Ranger

That damned fox; the devil himself could not catch him.
British Officer Banastre Tarleton about Francis The Swampfox Marion

It's easy to make it into the Regiment, but the hard part is maintaining the standard every single day.
75th Ranger Regiment Saying

Sua Sponte
Latin for of their own accord, motto of the 75th Ranger Regiment

De Oppresso Liber
Latin for To free the oppressed, motto of Special Forces

The guerilla must move amongst the people as a fish swims in the sea.
Mao Tse-Tung

The fundamental principle of revolutionary war: strike to win, strike only when success is certain; if not, then, don't strike.
GEN Vo Nguyen Giap

A standard question for a new man was why he volunteered for parachuting and whether he enjoyed it. On one occasion, a bright eyed recruit startled me by replying to the latter of the question with a resounding 'No, sir.' 'Why then, if you don't like jumping did you volunteer to be a paratrooper?' I asked. 'Sir, I like to be with people who do like to jump,' was the reply. I shook his hand vigorously and assured him that there were at least two of us of the same mind in the Division.
GEN Maxwell Taylor

Most have a job, some have a commitment.
Special Forces Recruiting Poster

Who dares wins.
Motto of the British SAS

Nous defions.
Latin for we dare, we defy, we threaten

A Ph.D. who can win a bar fight.
Wild Bill Donovan when asked to describe the ideal OSS Candidate

We were not afraid to make mistakes because we were not afraid to try things that had not been tried before.
Wild Bill Donovan

If you define leadership as having a vision for an organization, and the ability to attract, motivate and guide followers to fulfill that vision, you have Bill Donovan in spades.
Fisher Howe, special assistant to General Donovan

The OSS was an organization designed to do great things.
Professor E. Bruce Reynolds

My job entailed teaching men to kidnap, steal, cheat, or kill by the quickest, most ungentlemanly means possible.
COL Aaron Bank

American parachutists—Devils in Baggy Pants—are less than 100 meters from my outpost line. I can't sleep at night; they pop up from nowhere and we never know when or how they will strike next. Seems like the black-hearted devils are everywhere.

Dead German diary, Anzio 1944

DAS DICKE ENDE KOMMT NOCH! or The worst is yet to come

Motto of the 1st Special Service Force

You will be fighting an elite Canadian-American Force. They are treacherous, unmerciful and clever. You cannot afford to relax. The first Soldier or group of Soldiers capturing one of these men will be given a 10 day furlough.

Order found on Nazi prisoner at Anzio referring to the 1st Special Service Force

Humorous

It is difficult to be a good noncommissioned officer. If it had been easy, they would have given it to the officer corps.

SMA William A. Connelly

Logistics is tough. If it were easy it would be called operations.

A Tired BDE S-4

War is hell, but combat is a mother f__k_r.

COL Charlie Beckwith

Hard times don't last, but hard men do.

Motto 3rd SFG

When things go wrong in your command, start searching for the reason in increasingly larger concentric circles around your own desk.

GEN Bruce Clarke

It is a good thing for an uneducated man to read books of quotations
Prime Minister Winston Churchill: My Early Life (1930) ch. 9.

If everyone is thinking alike, someone isn't thinking.
General George Patton Jr

I thought to myself, Join the army. It's free. So I figured while I'm here I'll lose a few pounds.
John Candy in Stripes

The reason the American Army does so well in wartime, is that war is chaos, and the American Army practices it on a daily basis.
From a post-war debriefing of a German General

Remember, one grenade is nice, but two is always better.
CSM Winston Clough

I'm pacing myself, sergeant.
Bill Murray in Stripes

I ain't got time to bleed.
Jesse Ventura in Predator

Army officers are intelligent. Give them the bare tree, let them supply the leaves.
GEN George Marshall

You can always tell an old Soldier by the inside of his holsters and cartridge boxes. The young men carry pistols and cartridges; the old ones, grub.
George Bernard Shaw

PowerPoint makes us stupid.
Marine General James Mattis

Be polite, be professional, but have a plan to kill everybody you meet.
General James Mattis

I come in peace, I didn't bring artillery. But I am pleading with you with tears in my eyes: If you f%^$ with me, I'll kill you all.
General James Mattis

I think war might be God's way of teaching us geography.
Paul Rodriguez

A prisoner of war is a man who tries to kill you and fails, and then asks you not to kill him.
Prime Minister Winston Churchill

I never trust a fighting man who doesn't smoke or drink.
ADM William Halsey

The number of medals on an officer's breast varies in inverse proportion to the square of the distance of his duties from the front line.
Charles Montague

I've got to go meet God- and explain all those men I killed at Alamein.
Field Marshall Viscount Montgomery

For my part I prefer fifty thousand rifles to fifty thousand votes.
Benito Mussolini

Charlie don't surf.
Robert Duvall, Apocalypse Now

Our country will, I believe, sooner forgive an officer for attacking an enemy than for letting him alone.
ADM Viscount Nelson

Anybody who doesn't have fear is an idiot. It's just that you must make the fear work for you. Hell, when somebody shot at me, it made me madder than hell, and all I wanted to do was shoot back.
GEN Robin Olds

Nothing concentrates the military mind so much as the discovery that you have walked into an ambush.
Thomas Packenham

A ship without Marines is like a garment without buttons.
ADM David Porter

Nuts!
GEN Anthony McAuliffe

Paperwork will ruin any military force.
Chesty Puller

Grunt- Term of affection used to denote that filthy, sweaty, dirt-encrusted, footsore, camoflauge-painted, ripped trousered, tired, sleepy, beautiful little son of a b--- who has kept the wolf away from the door for over two hundred years.
H.G. Duncan

I can't say I was ever lost, but I was bewildered once for three days.
Daniel Boone

If you think you can, or if you think you can't…either way you're right.
Henry Ford

In times like these, it helps to recall that there have always been times like these.
Paul Harvey

If you can't change your circumstances, change your perspective.
Frank W. Davis

All it takes is one 'oh sh#$' to knock out a thousand 'atta boys.'
Ricky Reeves

If you want to give a man credit, put it in writing. If you want to give him hell, do it on the phone.
Lee Iacocca

Nothing endures but change.
Heraclitus

Next week there can't be any crisis. My schedule is already full.
Henry Kissinger

If you think you are leading and turn around to see no one following, then you are just taking a walk.
Benjamin Hooks

Artillerymen believe the world consists of two types of people; other Artillerymen and targets.
Unknown

One bad general is worth two good ones.
Napoleon Bonaparte

We are not retreating, we are advancing in another direction.
GEN Douglas MacArthur

Before a war military science seems a real science, like astronomy; but after a war it seems more like astrology.
Rebecca West

The pen may be stronger than the sword... but I'd rather have a sword in a dark alley.
Andrew Warnick

In this country we find it pays to shoot an admiral from time to time to encourage the others.
Voltaire

If your sword is too short, take one step forward.
ADM Marquis Heihachiro

The bullet is a mad thing; only the bayonet knows what it is about.
Field Marshall Prince Aleksandr V

The shovel is the brother to the gun.
Carl Sandburg

I too dabbled in pacifism once, not in 'Nam of course.
John Goodman in The Big Lebowski

You can't say that civilization don't advance, for in every war they kill you a new way.
Will Rogers

Diplomats are just as essential to starting a war as Soldiers are for finishing it…You take diplomacy out of war, and the thing would fall flat in a week.
Will Rogers

Find the enemy and shoot him down. Anything else is nonsense.
CPT Manfred von Richthofen (The Red Baron)

The person who knows 'how' will always have a job. The person who knows 'why' will always be his boss.
Diane Ravitch

Nothing concentrates the military mind so much as the discovery that you have walked into an ambush.
Thomas Packenham

Mr. President, I'm finer than the hair on a frog's back.
GEN Tommy Franks

People get the history they deserve.
GEN Charles de Gaulle

I really wish you were my First Sergeant, but I've already had a couple ones.
Pauly Shore, In the Army Now

They couldn't hit an elephant at this dist...
MG John Sedgwick, killed by a sniper at Spotsylvania 1864

Hurrah, boys, we've got them!
GEN George Armstrong Custer

Everyone's a pacifist between wars. It's like being a vegetarian between meals.
Colman McCarthy

The major force in world history is sheer dumbness.
Eric Wolf

Only the winners decide what were war crimes.
Gary Wills

The military don't start wars. Politicians start wars.
GEN William Westmoreland

Fortunate is the general staff which sees a war fought the way it intends.
Richard Watt

Scriptures

Blessed are the peacemakers: for they shall be called the children of God.

Matthew 5:9

The LORD is the one who goes ahead of you; He will be with you. He will not fail you or forsake you. Do not fear or be dismayed.

Deuteronomy 31:8

And I heard the voice of the Lord saying, Whom shall I send? And who will go for us? And I said Here I am. Send me.

Isaiah 6:8

Do not be afraid of the terrors of the night, nor the arrow that flies in the day.

Psalm 91:5

Greater love hath no man than this, that a man lay down his life for his friends.

John 15:13

God is our refuge and our strength, a very present help in trouble.
Psalm 46:1

In God's hands I have put my trust; I will not be afraid what man can do unto me.
Psalm 56:11

For with God nothing shall be impossible.
Luke 1:37

When a strong man, fully armed, guards his house, his possessions are safe.
Luke 11:21

Praise be to the LORD my Rock, who trains my hands for war, my fingers for battle.
Psalm 144:1

But those who hope in the LORD will renew their strength. They will soar on wings like eagles; they will run and not grow weary, they will walk and not be faint.
Isaiah 40:31

Be strong and courageous. Do not be afraid or terrified because of them, for the LORD your God goes with you; he will never leave you nor forsake you.
Deuteronomy 31:6

If you have faith as a mustard seed, you will say to this mountain, Move from here to there,' and it will move; and nothing will be impossible for you.
Matthew 17:20
'

By their fruits ye shall know them.
Matthew 7:20

Blessed are they who maintain justice, who constantly do what is right.
Psalm 106:3

As the Philistine moved closer to attack him, David ran quickly toward the battle line to meet him. Reaching into his bag and taking out a stone, he slung it and struck the Philistine in the forehead... And thus David triumphed over Goliath with a sling and a stone.
1 Samuel 17:48-50

The LORD is a man of war: The LORD is his name.
Exodus 15:3

How are the mighty fallen in the midst of the battle! O Jonathan, thou wast slain in thine high places.
2 Samuel 1:25

Infamous Leaders

The Iraqi people are capable of fighting to the victorious end which God wants... the blood of our martyrs will burn you!

Saddam Hussein, August 1990

Victory will be ours soon, Iraqis will strike the necks as God has commanded you.

Saddam Hussein

Iraqis have taught Bush a lesson that turned his concept upside down.

Iraq's al-Thawra newspaper

I ask you, being an Iraqi person, that if you reach a verdict of death, execution, remember that I am a military man and should be killed by firing squad and not by hanging as a common criminal.

Saddam Hussein At his trial for genocide and crimes against humanity, July 2006

My name is Saddam Hussein. I am the president of Iraq, and I want to negotiate.

Saddam Hussein to US troops who captured him in a hole in the ground near Tikrit

Yours is a society which cannot accept 10,000 dead in one battle.

Saddam Hussein

When the enemy starts a large scale battle, he must realize that the battle between us will be open wherever there is sky, land and water in the entire world... There are no weapons of mass destruction in Iraq. Well, give us time and the necessary means, and we will produce any weapon they want, and then we will invite them to come and destroy them.

Saddam Hussein

God is on our side, and Satan is on the side of the United States.

Saddam Hussein

We are not intimidated by the size of the armies, or the type of hardware the US has brought.
Saddam Hussein

Don't provoke a snake unless you have the intention and power to cut off its head.
Saddam Hussein

We did not find it difficult to deal with Bush and his administration, because it is similar to regimes in our countries both types include many who are full of arrogance and greed.
Osama bin Laden

I have sworn to only live free. Even if I find bitter the taste of death, I don't want to die humiliated or deceived.
Osama bin Laden

I'm fighting so I can die a martyr and go to heaven to meet God. Our fight now is against the Americans.
Osama bin Laden

We love death. The U.S. loves life. That is the difference between us two.
Osama bin Laden

We treat them in the same way. Those who kill our women and innocent, we kill their women and innocent, until they refrain.
Osama bin Laden

How fortunate for governments that the people they administer don't think.
Adolf Hitler

The leader of genius must have the ability to make different opponents appear as if they belonged to one category.
Adolf Hitler

The great masses of the people will more easily fall victims to a big lie than to a small one.
Adolf Hitler

Universal education is the most corroding and disintegrating poison that liberalism has ever invented for its own destruction.
Adolf Hitler

Success is the sole earthly judge of right and wrong.
Adolf Hitler

Who says I am not under the special protection of God?
Adolf Hitler

The day of individual happiness has passed.
Adolf Hitler

He alone, who owns the youth, gains the future.
Adolf Hitler

I do not see why man should not be just as cruel as nature.
Adolf Hitler

Humanitarianism is the expression of stupidity and cowardice.
Adolf Hitler

It is always more difficult to fight against faith than against knowledge.
Adolf Hitler

If you tell a big enough lie and tell it frequently enough, it will be believed.
Adolf Hitler

As a Christian I have no duty to allow myself to be cheated, but I have the duty to be a fighter for truth and justice.
Adolf Hitler

You can get much farther with a kind word and a gun than you can with a kind word alone.
Al Capone

Now I know why tigers eat their young.
Al Capone

I have built my organization upon fear.
Al Capone

To sum it all up, I must say that I regret nothing.
Adolf Eichmann

I was one of the many horses pulling the wagon and couldn't escape left or right because of the will of the driver.
Adolf Eichmann

Shoot first and ask questions later, and don't worry, no matter what happens, I will protect you.
Hermann Goering

Would you rather have butter or guns? Preparedness makes us powerful. Butter merely makes us fat.
Hermann Goering

Education is dangerous   every educated person is a future enemy.
Hermann Goering

Whenever I hear the word culture, I reach for my Browning!
Hermann Goering

American soldiers must be turned into lambs and eating them is tolerated.
Muammar Qaddafi

I am sailing out along parallel 32.5 to stress that this is the Libyan border. This is the line of death where we shall stand and fight with our backs to the wall. (On planning confrontation with US Sixth Fleet in Mediterranean)
Muammar Qaddafi

If Abu Nidal is a terrorist, then so is George Washington. (Reply to President Ronald Reagan in defense of Palestinian terrorist)
Muammar Qaddafi

The man who carried out the attack is still in power and still insane, so we shall expect another attack any minute. (On President Ronald Reagan)
Muammar Qaddafi

We are capable of destroying America and breaking its nose.
Muammar Qaddafi

There are no American infidels in Baghdad. Never!
Iraqi Information Minister Muhammed Saeed alSahaf

My feelings as usual we will slaughter them all
Iraqi Information Minister Muhammed Saeed alSahaf

Our initial assessment is that they will all die
Iraqi Information Minister Muhammed Saeed alSahaf

> I blame Al Jazeera. They are marketing for the Americans!
>
> Iraqi Information Minister Muhammed Saeed alSahaf

> God will roast their stomachs in hell at the hands of Iraqis.
>
> Iraqi Information Minister Muhammed Saeed alSahaf

> They're coming to surrender or be burned in their tanks.
>
> Iraqi Information Minister Muhammed Saeed alSahaf

> Death is the solution to all problems. No man—no problem.
>
> Joseph Stalin

> Americans are the great Satan, the wounded snake.
>
> Ayatollah Khomeini

> To read too many books is harmful.
>
> Mao Zedong

You cannot run faster than a bullet.
Idi Amin

I want you to know that everything I did, I did for my country.
Pol Pot

A man who dreads trials and difficulties cannot become a revolutionary. If he is to become a revolutionary with an indomitable fighting spirit, he must be tempered in the arduous struggle from his youth. As the saying goes, early training means more than late earning.
Kim Jong il

Equality means nothing unless incorporated into the institutions.
Slobodan Milosevic

I didn't say 'former president,' I said 'president,' and I have the constitutional rights according to the constitution, including immunity from prosecution.
Saddam Hussein

Ideas are more powerful than guns. We would not let our enemies have guns, why should we let them have ideas.
Josef Stalin

One death is a tragedy; one million is a statistic.
Josef Stalin

One man's terrorist is another man's freedom fighter
first written by Gerald Seymour in his 1975 book Harry's Game

Whoever stands by a just cause cannot possibly be called a terrorist
Yasser Arafat

I come bearing an olive branch in one hand, and the freedom fighter's gun in the other. Do not let the olive branch fall from my hand.
Yasser Arafat

Continue to press on, Soldiers of freedom! We will not bend or fail until the blood of every last Jew from the youngest child to the oldest elder is spilt to redeem or land!
Yasser Arafat

This is my homeland no one can kick me out.
Yasser Arafat

Anyone who doesn't regret the passing of the Soviet Union has no heart. Anyone who wants it restored has no brains.
Vladimir Putin

In the Soviet army it takes more courage to retreat than advance.
Joseph Stalin

The soviet people want full-blooded and unconditional democracy.
Mikhail Gorbachev

It seems that the most important thing about Reagan was his anti Communism and his reputation as a hawk who saw the Soviet Union as an 'evil empire.'
Mikhail Gorbachev

Jesus was the first socialist, the first to seek a better life for mankind.
Mikhail Gorbachev

What we need is Star Peace and not Star Wars.
Mikhail Gorbachev

Certain people in the United States are driving nails into this structure of our relationship, then cutting off the heads. So the Soviets must use their teeth to pull them out.
Mikhail Gorbachev

If people don't like Marxism, they should blame the British Museum.
Mikhail Gorbachev

Whether you like it or not, history is on our side. We will bury you.
Nikita Khrushchev

America has been in existence for 150 years and this is the level she has reached. We have existed not quite 42 years and in another seven years we will be on the same level as America. When we catch you up, in passing you by, we will wave to you.
Nikita Khrushchev

I am a Marxist-Leninist and I will be one until the last day of my life.
Fidel Castro

Berlin is the testicles of the West. Every time I want the West to scream, I squeeze on Berlin.
Nikita Khrushchev, 1962

The survivors (of a nuclear war) would envy the dead.
Nikita Khrushchev, July 20, 1963

If you (the USA) start throwing hedgehogs under me, I shall throw a couple of porcupines under you.
Nikita Khrushchev, November 7, 1963

Capitalism is using its money; we socialists throw it away.
Fidel Castro, November 8, 1964

I began revolution with 82 men. If I had to do it again, I do it with 10 or 15 and absolute faith. It does not matter how small you are if you have faith and plan of action.
Fidel Castro

I find capitalism repugnant. It is filthy, it is gross, it is alienating... because it causes war, hypocrisy and competition.
Fidel Castro

A revolution is a struggle to the death between the future and the past.
Fidel Castro

No thieves, no traitors, no interventionists! This time the revolution is for real!
Fidel Castro

They talk about the failure of socialism but where is the success of capitalism in Africa, Asia and Latin America?
Fidel Castro

A revolution is not a bed of roses.
Fidel Castro

I am Fidel Castro and we have come to liberate Cuba.
Fidel Castro

The left is back, and it's the only path we have to get out of the spot to which the right has sunken us. Socialism builds and capitalism destroys.
Hugo Chavez

I am convinced that the path to a new, better and possible world is not capitalism, the path is socialism.
Hugo Chavez

I hereby accuse the North American empire of being the biggest menace to our planet.
Hugo Chavez

They do not walk in... the path of Christ.
Hugo Chavez

A lie told often enough becomes the truth.
Vladimir Lenin

One man with a gun can control 100 without one.
Vladimir Lenin

The way to crush the bourgeoisie is to grind them between the millstones of taxation and inflation.
Vladimir Lenin

When one makes a Revolution, one cannot mark time; one must always go forward or go back. He who now talks about the 'freedom of the press' goes backward, and halts our headlong course towards Socialism.
Vladimir Lenin

Give me four years to teach the children and the seed I have sown will never be uprooted.
Vladimir Lenin

Fascism is capitalism in decay.
Vladimir Lenin

Crime is a product of social excess.
Vladimir Lenin

Index

A

Abbott, Lyman 179
Abrams, GEN Creighton 10, 86
Adams, John Quincy 25
Adcock, F.E. 68
ADM William Halsey 166
Aesop 189
Africanus, Scipio 95
Akers, Albert 9
Aldrich, Thomas 111
Aldrin, Buzz 149
Aleghieri, Dante 42
Aleksandr, Field Marshall Prince 62, 64
alSahaf, Muhammed Saeed 239, 240
Amiel, Henri 46
Amin, Idi 241
Animator, Alexander 43
Arafat, Yasser 242, 243
Aristotle 7, 34, 57
Arsenault, Leo 2
Aurelius, Emperor Marcus 59, 102, 198
Autry, James 21, 190

B

Bach, C.A. 20
Baker, COL AJ 'Bo' 206
Bank, COL Aaron 212
Barry, Dave 117
Bartol, Cyrus A. 98
Barton, Bruce 38
Baruch, Bernard 53
Beckham, Larry 101
Beckwith, COL Charlie 205, 214

Beecher, Henry Ward 32
Bennis, Warren G. 2
Bernbach, William 183, 186
Bin, Sun 24, 171
Blair, Tony 121
Blanchard, Felix Doc 192
Boardman, George Dana 97
Bonaparte, Napoleon 7, 9, 29, 31, 39, 44, 48, 79, 86, 87, 93, 106, 109, 113, 119, 124, 151, 167, 196, 199, 222
Bonta, Stanley 9
Boone, Daniel 220
Bradley, GEN Omar 8, 10, 56, 80, 107, 146, 147, 164, 173
Braude, Jacob 43
Brault, Robert 175
Brecht, Bertolt 117
Bremer, Paul 144
Brown, H. Jackson 173
Browning, Elizabeth Barrett 22, 135
Brown, John Mason 43
Bruce, Wallace 137
Bryant, Paul Bear 184
Buchan, John 27
Burden, Virginia 188
Burke, Edmund 37
Burnett, Carrell 14
Burnham, William 110
Burns, H.M.S. 22
Bushnell, Horace 35
Bush, President George W. 128, 129, 130, 143
Buxton, Thomas 36

C

Caesar, Julius 111, 118
Candy, John 215
Canning, George 135
Capone, Al 236, 237
Carlyle, Thomas 35, 36

Carter, Jimmy 174
Carter, Rosalynn 2
Carty, Margaret 188
Carver, Field Marshall Lord 93, 122
Castro, Fidel 245, 246, 247
Chabrias 8
Chamberlain, COL Joshua 92, 144, 145
Chamberlain, MG Joshua 152
Chavez, Hugo 247, 248
Chesterfield, Lord 97
Churchill, Prime Minister Winston 34, 48, 55, 70, 79, 82, 83, 86, 88, 94, 96, 112, 113, 120, 122, 124, 127, 128, 134, 143, 159, 179, 197, 215, 217
Cicero 15, 18
Clarke, Arthur C. 31, 34
Clarke, GEN Bruce 11, 19, 53, 147, 214
Clark, Eric 13
Clark, James 173
Clausewitz, MG Karl von 3, 8, 75, 82, 86, 95, 103, 106, 110, 124, 161, 162
Clear, LTC Warren 79
Clemenceau, Georges 75
Clough, Bronston 3, 14
Clough, CSM Winston 111, 215
Cloward, Robert 178
Collins, LTG Arthur 62
Confucius 20, 56, 176
Connelly, SMA William A. 214
Conte, Joseph Le 97
Coolidge, Calvin 139
Corbett, James 37
Covey, Steven 21
Craig, GEN Malin 157
Craig, W. Marshall 40
Crane, Frank 172
Crane, Stephen 200
Crimson, Howard 42
Crockett, Davy 174
Cromwell, Oliver 54, 171

Cunningham, ADM Andrew 90
Custer, GEN George Armstrong 225

D

Daly, Gunnery Sergeant Dan 109
Darby, Col. William O. 68
Darnis, Geraud 176
Davis, Elmer 137
Davis, Frank W. 220
Decatur, ADM Stephen 142
DeGaulle, GEN Charles 99
Dickinson, John 136
Disraeli, Benjamin 38, 39
Donitz, ADM Karl 202
Donovan, Wild Bill 211, 212
Dostum, GEN 208
Downing, GEN Wayne 205
Downs, Hugh 47
Dragomirov, M.I. 87
Drake, Joseph 137
Drucker, Peter 184
Dumas, Alexandre 182
Duncan, H.G. 219
DuPree, Max 45
Durant, Will 48
Duvall, Robert 218

E

Edison, Thomas 46
Edwards, Tryon 149
Eichmann, Adolf 237
Eimes, Leroy 25
Einstein, Albert 25, 44, 45, 178
Eisenhower, GEN Dwight 4, 9, 49, 50, 80, 86, 112, 121, 133, 147, 171
Eliot, T.S. 117

Emerson, Ralph Waldo   26, 31, 35, 41, 111, 135, 141, 182, 186
Emmons, GEN Delos   194
Epstein, Joseph   116
Erfurth   166
Erskine, John   22
Euripides   81, 114
Ewald, LTG Johann von   49

F

Fehrenbach, T.R.   77
Field, Franklin   17
Fisher, British Sea Lord John   75, 85
Fish, Hamilton   137
Foch, Ferdinand   58, 67, 165, 168
Forbes, B.C.   20, 41
Forbes, Malcolm   165
Ford, Henry   182, 185, 220
Forgy, Chaplain Howell   79
Forrest, Nathan Bedford   165
Fosdick, Henry   81, 181
France, Anatole   192
Frances, Charles   194
Franklin, Benjamin   40, 44, 136, 184
Franks, GEN Fred   201
Franks, GEN Tommy   224
Fritsch, GEN Werner von   104
Fry, James C.   1
Fuller, MG J.F.C.   92, 94, 162
Fuller, Thomas   114

G

Gandhi, Mahatmas   172
Gardner, Bryan   20, 35
Garfunkel, Art   116
Garnier, SGT   83

Gates, Bill 1
Gaulle, GEN Charles de 105, 107, 224
Gavin, MG J.M. 82
Geikie, John C. 98
General George S. Patton 69
GEN George S. Patton 6, 157
George, David Lloyd 38
Gerecht, CSM(R) Mark 98
Giap, GEN Vo Nguyen 210
Gibbon, Edward 66
Gilder, Richard Watson 134
Glasgow, Arnold 15, 19, 20
Glover, GEN Sir james 107, 171
Goering, Hermann 237, 238
Goodman, John 223
Gorbachev, Mikhail 243, 244
Gordon, David 41
Gough, J. B. 98
Grahn, Dennis 205
Grant, GEN Ulysses S. 78, 94, 124, 138, 161, 165, 167, 179
Gray, GEN Alfred 78, 125
Gunther, John 138

H

Haig, Field Marshall Earl 127
Hale, Edward 43
Hale, Nathan 142
Halsey, ADM William 217
Hammarskjold, Daj 118
Hamm, Mia 189
Hannibal 45, 95
Harding, Amber 188
Hart, Gary 139
Hart, Sir Basil Liddell 92, 152, 156, 157, 163
Harvey, Paul 220
Heihachiro, ADM Marquis 222
Heinlein, Robert A. 73

Helvetius 148
Hemingway, Ernest 76, 195
Henry, Patrick 140
Heraclitus 68, 100, 109, 221
Herder, Johann Gottfried Von 32
Heroditus 69
Hewlett, William 182
Hindenburg, Field Marshall Paul von 156, 200
Hirschfield, Tom 165
Hitler, Adolf 79, 234, 235, 236
Homer 109, 144, 150, 153, 168, 182, 199
Hooks, Benjamin 221
Horrocks, GEN Sir Brian 86
Howe, Edgar Watson 55, 58, 180
Howe, Fisher 212
Howze, GEN Hamilton 198
Hubbard, Elbert 36
Hughes, BG Chris 53
Hughes, Charles Evans 175
Humphrey, Hubert 101
Hussein, Saddam 231, 232, 233, 241
Hutchinson, Virginia 36

I

Iacocca, Lee 13, 15, 183, 221
Ibraham, Anwar 19
Ieyasu, Tokugawa 156
Ikanga'a, Juma 57
il, Kim Jong 241
Ingersoll, Robert G. 136
Iskander, Kai Ka'us Ibn 126

J

Jackson, Andrew 94, 110, 122
Jackson, General Thomas Stonewall 33, 50, 53, 60, 72, 73, 76, 93, 103, 118, 143

James, William 56
Jefferson, Thomas 31, 100, 101, 131, 141
Jeffery, Raili 46
Jobs, Steve 3
Johnson, Samuel 33, 36, 45
Jomini, LTG Antoine-Henri Baron de 96
Jones, John Paul 67, 75, 100
Jong, Erica 117
Jordan, Michael 185

K

Keller, Helen 185
Kennedy, John F. 41, 131, 132, 139, 207, 208
Kennedy, Robert F. 114, 174
Kerrey, Bob 102
Kesselring, Field Marshall Albert 61
Khomeini, Ayatollah 240
Khrushchev, Nikita 245, 246
King, ADM Ernest 197
King, James 34
King, Martin Luther 51, 99, 134, 174
Kipling, Rudyard 39, 62
Kirkover, CSM Herbert 208
Kissinger, Henry 26, 221
Knight, Charles 16
Knox, J.S. 19
Koch, Dr. G.P. 50
Koesther, Arthur 78
Krulak, LTG Victor 158, 160
Kung, Ts'ao 115

L

Laden, Osama bin 89, 233, 234
Lasorda, Tommy 47
Lawrence, CPT James 144
Lawrence, T.E. 24, 65, 78, 161, 163, 209

Lazarus, Emma 141
Ledru-Rollin, Alexandre-Auguste 4
Leeb, Von 165
Lee, General Robert E. 28, 69, 104, 151, 154, 164, 177
Lejeune, MG John 23, 28, 61, 152
Lenin, Vladimir 248, 249
Letterman, David 116
Lewis, Christopher 57
Liang, Zhuge 159
Liddon, Henry Louis 58
Lincoln, Abraham 37, 43, 99, 105, 130, 140, 146, 158, 164, 174
Lippmann, Walter 4
Lloyd, Henry 78
Logan, John A. 136
Lombardi, Vince 37, 42, 55, 187
Luccock, H.E. 187
Ludendorff, GEN Erich 154
Luther, John 9

M

MacArthur, General Douglas 6, 30, 52, 70, 71, 76, 100, 108, 132, 141, 166, 167, 196, 222
Machiavelli, Niccolo 88
Mahan, Admiral Alfred Thayer 163
Mahoney, David 42
Malone, COL Dandridge 61
Malone, COL Mike 59
Maltke, Von 56
Markos, C. 172
Marshall, BG S.L.A. 14, 29, 64, 66, 80, 107, 125, 143, 153, 157, 159, 200, 201
Marshall, GEN George C. 1, 23, 88, 192, 193, 195, 197, 216
Mattis, General James 216, 217
Mauldin, SGT Bill 88
McAuliffe, GEN Anthony 219
McCarthy, Colman 81, 225

McCarthy, Cormac 85
McCrae, John 81
McGannon, Donald H. 19
McRaven, ADM William 207
Meade, Margaret 188
Meadows, Dick 207
Meinertzhagen, R.E. 17, 101
Menander 44
Meyer, GEN E. C. 73
Miller, Henry 7
Mill, John Stuart 74, 110
Milosevic, Slobodan 241
Mitchell, Margaret 125
Moliere 45
Moltke, Field Marshall Helmuth von 163
Montagne, Michel de 114
Montague, Charles Edward 82, 217
Montecuccoli, Field Marshall 84
Montgomery, Field Marshall Bernard 16, 17, 84, 157, 159, 195, 197, 198, 218
Moore, Mary Tyler 116
Moran, Lord 16
Mosby, COL John S. 96, 209
Motecuhzoma 126
Munenori, Yagyu 89
Mussolini, Benito 218

N

Naisbitt, John 25
Napier, GEN Sir William 92
Narosky, José 135
Nelson, ADM Viscount 93, 103, 218
Newman, Cardinal John Henry 40
Newman, MG Aubrey 84
Nietzsche, Friedrich 84
Nightingale, Florence 49
Nimitz, Admiral Chester 27, 109
Nixon, Richard 85, 91

Noble, Charles 46
Nobushige, Takeda 105

O

Oath, Athenian 140
O'Brien, Michael 194
Olds, GEN Robin 218
Oliver, James 41
Onassis, Jacqueline Kennedy 58
Ortberg, John 125
Orwell, George 25, 67
Osler, Sir William 39
Ovid 46, 50
Ozick, Cynthia 135

P

Pachacutec 122
Packenham, Thomas 219, 224
Pagonis, LTG William 83
Paine, Thomas 37, 97, 133
Parry, COL F.F. 83
Paterson, Katherine 116
Patton, General George S. 5, 14, 30, 52, 53, 54, 63, 64, 65, 70, 71, 72, 74, 76, 77, 87, 89, 90, 93, 94, 99, 108, 114, 121, 132, 145, 146, 151, 153, 155, 162, 191, 197, 200, 215
Peale, Norman 179
Pericles 115, 142
Perot, H. Ross 159
Perry, Commodore Oliver Hazard 78
Pershing, GEN Blackjack 166
Peter, Lawrence J. 12
Peters, LTC Ralph 89
Peterson, Wilfred 44
Peters, Tom 13, 22
Phillips, COL William 79
Phillips, Wendell 13

Phormio 65, 85
Picq, Ardant Du 54
Picq, COL Charles Ardant du 151, 161, 191, 199
Picton, GEN Sir Thomas 84, 87
Pilar, GEN Gregario Del 115
Pitt, William 139
Plato 12, 37
Plautus 183
Plutarch 23
Pope, Alexander 176
Porter, ADM David 219
Pot, Pol 241
Powell, GEN Colin 24, 26, 55, 158
Puller, LTG Chesty 87, 201, 219
Purdy, CSM Donald 206
Putin, Vladimir 243
Putnam, Israel 128
Pyle, Ernie 88
Pyrrhus 150

Q

Qaddafi, Muammar 238, 239
Quarles, Francis 38, 40
Quincy, Josiah 40

R

Raeder, Grand ADM Erich 158
Ravitch, Diane 224
Ray, John 36
Reagan, Ronald 102, 208
Redmoon, Ambrose 117
Reeves, Ricky 220
Renatus, Flavius Vegetius 52, 54, 66, 126
Reynolds, Professor E. Bruce 212
Richter 180
Richthofen, CPT Manfred von 224

Ridgway, GEN Matthew 63, 104, 121
Robbins, Anthony 45, 51
Rockefeller, John 12
Rodriguez, Paul 217
Rogers, Will 47, 223
Rommel, Erwin 3, 51, 164, 169, 174
Roosevelt, Eleanor 119, 195
Roosevelt, Franklin D. 11, 123
Roosevelt, Teddy 18, 26, 27, 33, 99, 106, 112, 166
Rosewarne, VA 123
Rowe, COL Nick 206
Rumsfeld, Donald 59
Rumyantsev, GEN Piotr 170
Ruskin, John 186
Russ, Dr. Eric 180
Ruth, Babe 189

S

Saban, Nick 186
Safire, William 175
Saintsbury, Val 138
Sandburg, Carl 223
Saxe, Field Marshall Maurice Comte de 65, 151
Saying, Athenian 119
Schofield, Major General John M. 148
Schuller, Robert H. 59
Schwarzkopf, General H. Norman 6, 16, 24, 71, 77, 105, 204
Schweitzer, Dr. Albert 15, 179
Scotford, John 15
Scott, Sir Walter 118
Sedgwick, MG John 225
Selfridge, H. Gordon 19
Seneca 123
Seymour, Gerald 242
Shain, Merle 59
Shakespeare, William 30, 39, 69, 110, 136, 175
Shaw, George Bernard 216

Sherman, GEN William T.   69, 85, 91, 125, 144, 156, 201
Shidle, Norman   187
Shore, Pauly   225
Sickel   27
Sidney, Sir Philip   122
Sills, Beverly   38
Sill, Sterling   57
Simons, COL Arthur Bull   207
Simpson, Alan   175
Simpson, Louis   54
Slim, Field Marshall Viscount William   16, 17, 120, 150, 196, 198
Solomon   18
Solon   49
Sprague, William B.   32
Stalin, Josef   242
Stalin, Joseph   240, 243
Stambaugh, A.A.   22
Stark, BG John   126
Stengel, Casey   189
Stockdale, ADM James   18, 24, 172
Strong, COL Vincent   92
Summerall, Charles P.   11
Suvorov, Alexander V.   66
Syrus, Plubilius   183

T

Tacitus   4, 27
Tarleton, Banastre   209
Taylor, COL George   91
Taylor, GEN Maxwell   13, 14, 202, 211
Tennyson   118
Terence, Publius   108
Thoreau, Henry David   47, 97
Thucydides   52, 60, 70, 108, 112, 119, 153
Tilden, Samuel   190
Tiorio   15, 192
Tolstoy   193

Truman, Harry S. 1, 23, 110, 144
Truscott, BG Lucian 209
Tse, Chuang 183
Tse-Tung, Mao 154, 170, 210
Tucumsia 190
Tunis, GEN Viscount Alexander of 155
Turenne, Henri 4
Tutu, Archbishop Desmond 115
Twain, Mark 34, 173
Tzu, Lao 7
Tzu, Sun 5, 8, 29, 51, 74, 75, 83, 120, 121, 148, 162, 168, 169

V

Vegetius 58, 112, 113
Ventura, Jesse 216
Verity, George 184
V, Field Marshall Prince Aleksandr 60, 120, 223
Vilabla, Jose 90
Voltaire 32, 222

W

Walker, Greg 203
Wansong, Monk 176
Ware, Eugene F. 38
Warner, Brian 122
Warnick, Andrew 222
Washington, GEN George 54, 103, 139, 143, 147, 150, 154, 172, 176, 200
Watt, Richard M. 169, 226
Waugh, SGM Billy 206
Wavell, Field Marshall Viscount Archibald 63, 66, 81, 104, 162, 164, 168
Wayne, John 111, 142
Webb, James 201
Webster, Daniel 137, 140

Webster, John P. 181
Wesley, John 180
Westmoreland, GEN William 114, 226
West, Rebecca 222
Weyand, GEN Fred 134
Whitman, Walt 98
Wickham, GEN John 100
Wilde, Oscar 26
Willkie, Wendell 18
Wills, Gary 225
Wilson, H.H. 115
Wilson, John 192
Wilson, Woodrow 138, 184
Winters, Dick 119
Wolf, Eric 225
Wolffsohn, David 21
Woods, Tiger 205

X

Xenophon 29, 75, 89, 123, 150, 195

Z

Zedong, Mao 240
Zerhouni, Elias 23